# AMERICA'S RESPONSE TO TERRORISM

Brian Michael Jenkins and John Paul Godges

EDITORS

James Dobbins, Arturo Muñoz, Seth G. Jones, Frederic Wehrey,
Angel Rabasa, Eric V. Larson, Christopher Paul, Kim Cragin,
Todd C. Helmus, Brian A. Jackson, K. Jack Riley,
Gregory F. Treverton, Jeanne S. Ringel, Jeffrey Wasserman,
Lloyd Dixon, Fred Kipperman, and Robert T. Reville

CONTRIBUTORS

Funding for this book was made possible by RAND's Investment in People and Ideas program, which combines philanthropic contributions from individuals, foundations, and private-sector firms with earnings from RAND's endowment and operations to support innovative research on issues crucial to the policy debate but that reach beyond the boundaries of traditional client sponsorship.

**Library of Congress Cataloging-in-Publication Data**

The long shadow of 9/11 : America's response to terrorism / Brian Michael Jenkins, John Paul Godges, editors.
p. cm.
Includes bibliographical references.
ISBN 978-0-8330-5833-1 (pbk. : alk. paper)
1. Terrorism—United States—Prevention. 2. War on Terrorism, 2001-2009.
3. United States—Military policy—21st century. I. Jenkins, Brian Michael. II. Godges, John Paul.

HV6432.L64 2011
363.325'160973—dc23

2011026477



*Cover design by Peter Soriano.*



Published 2011 by the RAND Corporation
1776 Main Street, P.O. Box 2138, Santa Monica, CA 90407-2138
1200 South Hayes Street, Arlington, VA 22202-5050
4570 Fifth Avenue, Suite 600, Pittsburgh, PA 15213-2665
RAND URL: http://www.rand.org/
To order RAND documents or to obtain additional information, contact
Distribution Services: Telephone: (310) 451-7002;
Fax: (310) 451-6915; Email: order@rand.org

# Foreword

In remarks at the opening of the 14th NATO Review Meeting in Berlin, Germany, on September 19, 2001, just eight days after the 9/11 terrorist attacks, I characterized the terrorists as "a party to a virtual civil war within Islam—a war between extremists and moderates." I also noted that the civil war was not limited to Islam. Militant white supremacists in the United States often refer to Christian scripture to justify their acts. There are also extremist Jews, some of whom are violent. There are Hindu terrorists as well.

This means that our policies "should not focus on religion per se," I argued. "True adherents are not violent and do not support violence. The extremists who are our problem are fanatics. They are the antithesis of the religion they purport to represent."

Fortunately, the United States and its European allies have resisted making religion a focal point of counterterrorism efforts. As discussed in this volume, other U.S. policies in the Middle East since 9/11 have weakened and sometimes strengthened the hand of the extremists. We've had many policy successes and dealt the extremists many blows, but the Iraq War, the abuses at Abu Ghraib prison, and the indefinite detention of enemy combatants at Guantanamo Bay Naval Base have assisted the extremists in recruiting foot soldiers for their war.

In Berlin, I called for a policy that would balance offensive and defensive components. The offensive components would be international cooperation, diplomacy, intelligence, efforts to promote democracy and to silence hate mongering, police work, development assistance, counterproliferation, and military power. The defensive components would be passive as well as active, from transportation security and infrastructure protection to missile defenses and other military measures to deny enemy access to U.S. territories.

This volume can be considered a partial update of how well the United States has pursued many of these priorities. Its chapters track the ongoing civil war within Islam, assess crucial lessons from the wars in Afghanistan and Iraq, acknowledge how America has overreacted at times, identify opportunities that have yet to be explored, and discuss how America can continue to be true to its highest self.

More than a policy primer, this book turns a constructively critical eye toward America itself in the years since 9/11. The authors offer informed commentary on the larger social, cultural, military, and other implications of U.S. policymaking, while still remaining grounded in a solid research foundation. The hope is that these commentaries will lend a uniquely broad and farsighted perspective to the national dialogue on the legacy of 9/11.

James A. Thomson
President and Chief Executive Officer
RAND Corporation

*June 2011*

# Contents

**Foreword** ... iii
**Acknowledgments** ... xi

**Introduction: The Shadow of 9/11 Across America**
*Brian Michael Jenkins and John Paul Godges* ... 1
A Moment to Reflect ... 2
An Honest Accounting ... 4
An American Perspective ... 6
A Better Criterion ... 7

**PART ONE: HUMBLED BY HUBRIS** ... 9

**CHAPTER ONE**
**The Costs of Overreaction**
*James Dobbins* ... 15
Overreactions and Underestimates ... 15
Bruised but Not Broken ... 18
Related Reading ... 20

**CHAPTER TWO**
**A Long-Overdue Adaptation to the Afghan Environment**
*Arturo Muñoz* ... 23
How the Western-Lead Model Has Caused Problems ... 25
What an Afghan-Lead Model Could Have Looked Like ... 28
Related Reading ... 34

CHAPTER THREE
**Lessons from the Tribal Areas**
*Seth G. Jones* .... 37
Debating the Threat .... 38
Countering al Qaeda and Its Allies .... 42
A Long War .... 44
Related Reading .... 45

CHAPTER FOUR
**The Iraq War: Strategic Overreach by America—and Also al Qaeda**
*Frederic Wehrey* .... 47
The Narrative Godsend to al Qaeda .... 48
The University of Jihad? .... 49
The (Mis)management of Savagery .... 50
Lessons Learned by al Qaeda . . . .... 51
. . . and the United States .... 52
The Iraq War Versus the Arab Uprisings .... 53
Related Reading .... 54

**PART TWO: HOPEFUL AMID EXTREME IDEOLOGIES AND INTENSE FEARS** .... 57

CHAPTER FIVE
**Where Are We in the "War of Ideas"?**
*Angel Rabasa* .... 61
From the Generic to the Specific .... 62
Whose War Is It? .... 67
Related Reading .... 70

CHAPTER SIX
**Al Qaeda's Propaganda: A Shifting Battlefield**
*Eric V. Larson* .... 71
Al Qaeda Under Attack .... 74
Defending the Movement .... 78
Implications for U.S. Strategy and Policy .... 82
Related Reading .... 85

CHAPTER SEVEN
**Have We Succumbed to Nuclear Terror?**
*Brian Michael Jenkins* .......... 87
A Psychological Triumph .......... 88
Ambitions, Not Capabilities .......... 90
Absence of Warning .......... 92
Hypothetical Possession, Vivid Consequences .......... 93
Spinning Nuclear Fantasies .......... 95
Driven by Our Imagination .......... 96
Harnessing the Power .......... 97
Related Reading .......... 99

**PART THREE: TORN BETWEEN PHYSICAL BATTLES AND MORAL CONFLICTS** .......... 101

CHAPTER EIGHT
**Winning Every Battle but Losing the War Against Terrorists and Insurgents**
*Christopher Paul* .......... 105
Diminishing Support .......... 107
Winning the Battles *and* the Wars .......... 109
Related Reading .......... 110

CHAPTER NINE
**The Strategic Dilemma of Terrorist Havens Calls for Their Isolation, Not Elimination**
*Kim Cragin* .......... 113
The Ongoing Challenge .......... 115
Time for a Different Approach .......... 117
Related Reading .......... 120

CHAPTER TEN
**Our Own Behavior Can Be Our Weakest Link—or Our Strongest Weapon**
*Todd C. Helmus* .......... 121
Showing, Not Just Telling .......... 122
Shooting Oneself .......... 123

Counter Resentment, Counter Terrorism ..... 125
Related Reading ..... 127

**PART FOUR: DRIVEN BY UNREASONABLE DEMANDS** ..... 129

CHAPTER ELEVEN
**Don't Let Short-Term Urgency Undermine a Long-Term Security Strategy**
*Brian A. Jackson* ..... 133
Demanding Perfection Makes Failure Inevitable ..... 138
Sustainable Security Equals Success ..... 141
Time to Reduce the Pendulum Swings in Security Policy ..... 143
Related Reading ..... 145

CHAPTER TWELVE
**Flight of Fancy? Air Passenger Security Since 9/11**
*K. Jack Riley* ..... 147
Air Travel Is Safe and Secure ..... 150
Missed Opportunities ..... 153
The Decade Ahead ..... 158
Related Reading ..... 160

CHAPTER THIRTEEN
**The Intelligence of Counterterrorism**
*Gregory F. Treverton* ..... 161
From Nation-States to Terrorist Targets ..... 162
How Are We Doing? ..... 164
Related Reading ..... 167

**PART FIVE: INSPIRED TO BUILD A STRONGER AMERICA** ..... 169

CHAPTER FOURTEEN
**The Public Health System in the Wake of 9/11: Progress Made and Challenges Remaining**
*Jeanne S. Ringel and Jeffrey Wasserman* ..... 173
Progress Has Been Made on Many Fronts ..... 176
Important Challenges Remain ..... 178

A Path Forward .......... 181
Related Reading .......... 184

CHAPTER FIFTEEN
**The Link Between National Security and Compensation for Terrorism Losses**
*Lloyd Dixon, Fred Kipperman, and Robert T. Reville* .......... 185
Compensation for 9/11 Victims .......... 187
Current Compensation Mechanisms for Terrorist Attacks .......... 189
Designing a Strategy for the Future .......... 191
Related Reading .......... 193

CHAPTER SIXTEEN
**The Land of the Fearful, or the Home of the Brave?**
*Brian Michael Jenkins* .......... 195
A New Era Defined .......... 196
Alarms but No More 9/11s .......... 197
The Constitution Holds, Mostly .......... 198
Little Tolerance for Risk .......... 201
A Nation Continually at War .......... 203
American Values Damaged .......... 205
From External to Internal Threats .......... 205
Home of the Brave .......... 207
Related Reading .......... 207

**About the Editors** .......... 209

# Acknowledgments

In this challenging economic time for many institutions, the editors are particularly grateful to James A. Thomson, president and chief executive officer of the RAND Corporation, and Karen Gardela Treverton, special assistant to the president's office, for securing financial support for this book from RAND's Investment in People and Ideas program. Two additional RAND Corporation managers played central roles in supporting this project, from conceptualization to implementation: Jack Riley, vice president and director of the RAND National Security Research Division, and Andrew Morral, director of the RAND Homeland Security and Defense Center.

More than a score of reviewers offered astute criticisms that improved the content of this book. Those who reviewed the entire draft were Ambassador L. Paul Bremer III, the Honorable Richard J. Danzig, and Suzanne E. Spaulding.

Bremer, who served as chairman of the National Commission on Terrorism and later as head of the Coalition Provisional Authority in Iraq, is president and chief executive officer of World T.E.A.M. Sports, which organizes athletic events for disabled citizens. Danzig, former U.S. Secretary of the Navy, is chairman of the Center for a New American Security and a member of the Defense Policy Board, an advisory committee to the U.S. Department of Defense. Spaulding, an authority on U.S. national security and homeland security, was the executive director of both the National Commission on Terrorism and the Commission to Assess the Organization of the Federal Government to Combat the Proliferation of Weapons of Mass Destruction.

Lynn Karoly, director of research quality assurance at the RAND Corporation, held the review process to the highest standards while expediting matters when possible to help meet tight deadlines. Additional managers of the peer review process within the RAND Corporation were Andrew Morral, James Dobbins, Katharine Webb, Lisa Meredith, and John Parachini.

Stephen Watts, a RAND political scientist who reviewed and helped revise six chapters of the book, deserves special recognition for bearing this exceptional workload and for doing so with tireless goodwill. The other RAND peer reviewers were Charles Ries, Barbara Sude, Paul Davis, David Frelinger, Henry Willis, Tom LaTourrette, Admiral Robert Murrett, Mark Monahan, Arthur Kellermann, James Dertouzos, and Roger Molander.

Nancy Pollock and Donna Thomas, of the RAND National Security Research Division, volunteered their administrative expertise. RAND librarian Anita Szafran compiled a vast database of published RAND reports on terrorism and homeland security since 9/11 to help ensure that no major topic was overlooked.

Editor Janet DeLand gave the text its polish. Designer Peter Soriano gave the cover its gloss. The RAND Publications Department team of Jane Ryan, Paul Murphy, Stephan Kistler, Todd Duft, James Torr, Kelly Schwartz, Eileen La Russo, and John Warren shepherded the book through its production, proofreading, printing, marketing, and distribution. And Jennifer Gould, the director of strategic communications for the RAND Office of External Affairs, choreographed and coordinated the wide-ranging RAND outreach effort to bring this book to the attention of key policy audiences.

# Introduction: The Shadow of 9/11 Across America

*Brian Michael Jenkins and John Paul Godges*

*Brian Michael Jenkins, senior adviser to the president of the RAND Corporation, initiated RAND's research on terrorism in 1972. John Paul Godges is editor-in-chief of* RAND Review, *the flagship magazine of the RAND Corporation.*

It is, at this moment, nearly ten years since 9/11. The deadliest attacks in the annals of terrorism and the cause of the greatest bloodshed on American soil since the Civil War, the 9/11 attacks provoked the invasion of Afghanistan, which has become America's longest war. The attacks also prompted America's global campaign against terrorists and terrorism—a campaign that soon broadened to include the invasion of Iraq, a fundamental reorganization of the intelligence community, and a continuing national preoccupation with domestic security marked by the creation of a new national apparatus, the U.S. Department of Homeland Security, dedicated to the protection of American citizens against terrorist attacks.

The death in May 2011 of Osama bin Laden, founder and leader of al Qaeda, who declared war on the United States in 1996 and who was the driving force behind the 9/11 attacks, would seem to bracket, if not the war on terrorism, at least an important chapter in that war. While the killing of bin Laden led to a brief display of national euphoria, few analysts—and none of the authors in this volume—believe that his death spells the end of al Qaeda or its terrorist campaign. His demise is a semicolon in the ongoing contest, not a period.

Al Qaeda's future trajectory is not yet discernible. The organization has warned of retaliation and is under pressure to demonstrate to its foes and, more importantly, to its followers that the global terrorist enterprise inspired by bin Laden is still in business. With no other al Qaeda leader possessing bin Laden's symbolic authority, demonstrating prowess as a terrorist commander is one way of asserting leadership. Further terrorist attacks must be anticipated. As the tenth anniversary of 9/11 approaches, apprehension will increase. But the threat will persist for many years.

Much, of course, has changed in the past ten years. As in any war of long duration, there have been surprises—some of America's own making, such as its invasion of Iraq, followed by al Qaeda's own unmaking in the brutal and indiscriminate terrorist campaign it launched in response to the American occupation of that country. Al Qaeda's wanton slaughter of Muslims in Iraq and elsewhere provoked a powerful backlash among Muslims worldwide. The global economic crisis, America's financial difficulties, and the unpredicted popular uprisings in North Africa and the Middle East, although not directly connected with the war on terrorism, will nonetheless have a great impact on future American counterterrorist policy and strategy.

## A Moment to Reflect

Even before bin Laden's death, the tenth anniversary of America's response to 9/11 seemed an appropriate time for a thoughtful review of progress and future strategy. The perspective of a decade would reveal broad trends not apparent in shorter time frames. When we, the editors of this volume, first discussed the idea of such a review with RAND's management and staff, we made it clear that we did not want just a tenth-year anniversary anthology, a mere sampler of past RAND research. We wanted the participants in this project to not only draw upon their accumulated expertise and accrued knowledge but also go beyond what they had already published and reflect upon broader issues.

Did America as a nation overreact to 9/11? What did America do right? What did the country get wrong? Have there been lost opportunities or unwise approaches? What lessons have been learned? What might the country now do differently? What can Americans realistically expect from security? Has 9/11 changed how Americans view war? And has 9/11 changed *us* as Americans?

The contributing authors did not disappoint. Their essays are agile, yet muscular, recognizing the progress made in some areas but offering criticism where it is needed.

These are not the laments of insulated academics, dovish dons, or adherents of the kumbaya school of counterterrorism. Almost all of the authors were involved in terrorism research years before 9/11. Several of them complement their research with decades of firsthand experience in the armed forces; in the Central Intelligence Agency; in the U.S. Departments of State, Justice, and Defense; or as advisers to military commanders in Iraq and Afghanistan. These authors have been involved in intelligence collection and analysis. They have been on the front lines of diplomacy. They have seen war.

The long shadow of 9/11 sometimes makes it difficult to recall what things were like before terrorists crashed hijacked airliners into New York skyscrapers and the Pentagon, killing thousands. The United States had been concerned about the growing phenomenon of terrorism since the late 1960s and had played a major role in international efforts to combat it. Terrorism escalated in the 1980s and 1990s as terrorists increasingly demonstrated their determination to kill in quantity and their willingness to kill indiscriminately. Terrorist attacks on American targets abroad had already provoked a military response on several occasions, but these were single actions.

Prior to 9/11, neither Washington nor the American public was psychologically or politically prepared to launch not simply retaliatory strikes but a continuing military campaign against a terrorist movement—to wage war on terrorism, whatever that meant. Without 9/11, it would have been hard to imagine the subsequent American response. And even given that response, few in 2001 anticipated that the effort would last so long or prove so costly.

## An Honest Accounting

There is consensus in this volume that the United States has accomplished a great deal in the past ten years. Al Qaeda's capacity for centrally directed, large-scale terrorist operations has been greatly reduced, if not eliminated entirely. Declaring victory and turning our back, however, would be dangerous.

The United States cannot prevent every terrorist attack, but it is much better equipped today to handle future terrorist threats. U.S. intelligence has shifted its priorities from nation-states to transnational actors and has reconfigured itself to meet the new threats. The intelligence operation that led to the successful raid on bin Laden's compound in Abbottabad, Pakistan, displayed this greatly increased effectiveness. Al Qaeda's ranks have been decimated, its capabilities degraded, not only as a result of U.S. intelligence, military, and Special Operations but also very much as a consequence of unprecedented international cooperation among the world's security services and law enforcement organizations.

The authors in this book do not flinch at the invasion of Afghanistan, the continuing use of military force to destroy al Qaeda, or current efforts against the Taliban. They do, however, criticize the invasion of Iraq on grounds that it diverted attention from Afghanistan and the pursuit of al Qaeda, that the United States sidelined even willing allies to pursue military missions largely on its own, that military operations in both Afghanistan and Iraq ended up being stretched thin, and that the requirements of both counterinsurgency campaigns were ignored until late in the campaigns. The United States has been forced to learn—and to relearn—a great deal the hard way, especially about counterinsurgency operations.

The authors remain skeptical of current U.S. efforts to build up a large *national* army and police force in Afghanistan without simultaneously building up *local* forces, which seem closer to that country's traditions. There is further dissatisfaction with the continuing failure to deliberately combat al Qaeda's ideology or to support those who can. These essays are not just primers on theology or cultural sensitivity; they are pragmatic arguments about how to succeed.

America's approach to al Qaeda has been focused on destroying the organization, not confronting its ideology. From the outset, preventing further terrorist attacks took precedence. America pounded on al Qaeda's operational capabilities, not its beliefs, which were largely dismissed as fanaticism. But as several authors point out here, military power alone does not suffice.

The authors reject the idea that wars can always be won quickly and with minimal resources. Circumstances can change. Stuff happens, and it cannot be ignored because it does not correspond to initial expectations. The wars in Iraq and Afghanistan, initially billed as likely to be short, inexpensive, even self-financing, have turned out to be long and costly in lives and treasure—costlier in the long run because the United States, for reasons of domestic politics or plain arrogance, held absurd beliefs about their probable duration, failed to prepare for the worst, or walked away prematurely and had to return to a deteriorating situation. The authors observe this same arrogance in the institutional resistance to relearning lessons buried long ago with bad memories of the Vietnam War, and in Americans' tendency to ignore their own history.

If anything, the authors might be labeled American "traditionalists." They accept wars but are wary of the hubris that comes with America's great military power. They believe that despite America's military capacity, wars must be preceded by diplomacy to build powerful military coalitions. Although coalitions are difficult to manage and can become unwieldy, they are preferable to the notion that Americans can do it better alone. America traditionally goes to war with allies.

The authors hold dear traditional American values, lamenting the country's departure from these values in its abuse of prisoners—the subject of a debate that some politicians seem determined to revive. The authors cleave to the traditional American values of courage and self-reliance, underscoring the fact that the terrorist threat has been exaggerated to our own detriment. Terrorism is a risk, as are a lot of other things. Major terrorist attacks can cause horrific casualties, create international and domestic crises, impose huge costs on the economy, and set off psychological reactions that can corrode democracy itself.

But Americans can reduce the effects of terror by their own reactions to such events.

The authors sometimes harp on America's foibles, but they also harbor great hopes for America's future. Some authors assert that America overreacted after 9/11, both abroad (in Iraq) and at home (at the airports). Domestic political rhetoric and a voracious news media, meanwhile, assailed the public with terrorist threats, many of them the dark speculations of professional doomsayers. Yet the authors also see opportunities for strengthening America that have arisen partly because of 9/11, from reinforcing our public health system to redesigning our laws to promote community solidarity in times when we most need to rely on each other.

## An American Perspective

This volume offers an explicitly American point of view. It was our intent to look critically at America's experience and performance. This reflects the fact that America has often chosen to go it alone, determined to run its own show, unwilling to be fettered by assistance from others, ignoring advice that did not accord with its own perspectives—hubris again. It was our mandate as authors and editors to comment upon that American experience.

Nonetheless, we are aware of the enormous cooperation and collaboration that have taken place with international allies, especially, as pointed out, in the areas of intelligence and law enforcement, but also on the battlefield. As the continuing terrorist threat inspired by al Qaeda's ideology becomes more diffuse, this collaboration will become even more critical.

Another area of continuing importance will be the prevention of domestic radicalization and recruitment of terrorists to al Qaeda's cause—so-called "homegrown terrorism." America historically has been successful in assimilating immigrant populations, and thus far, al Qaeda's exhortations to America's Muslim population have produced meager results. America can learn from the experience of European and other nations in this area, while being wary of emulating

Europe's national efforts. America's political traditions are very different, making interaction with ethnic and religious communities much more a matter for local authorities than for Washington.

The American people have watched the Arab Spring of 2011 with rapt attention. The United States cannot take credit for this phenomenon, but American ideals and communications technologies, if not always American policies, may have cultivated some of those sprouts of democracy in the Middle East. The Arab Spring has demonstrated the irrelevance of al Qaeda's ideology to the political future of the region, but it has also created a challenge for U.S. diplomacy. Whatever new governments ultimately may emerge, counterterrorism seems unlikely to top their agendas, and it cannot be the single focus of U.S. relations with these governments.

## A Better Criterion

Americans frequently ask, Are we safer now? The question betrays the perspective of a victim. The answer is probably yes, but surely that cannot be the sole criterion of progress. Instead, we might ask, Is America stronger now? Can the *country* defend itself against current and would-be foes? Can the *country* sustain the perpetual state of preparedness in which, it seems, we must live? Is *America* capable of achieving its objectives?

Assessment here is more difficult, beyond the reach of research, but the answer seems mixed. America is probably organizationally and militarily better prepared now. It has gained a better understanding of this new kind of adversary. Having survived 9/11, Americans may be better prepared psychologically to deal with another terrorist attack. But national strength derives from more than the accumulation of warriors and weapons and endurance. It encompasses public spirit. A civic spirit. A sense that everyone is in it together.

America's struggle against terrorism has been national in name only. Except for the heavy burden borne unequally by those in the military and their families, the conflict remains a distant reality show to the rest of society. Conspicuous displays of patriotism disguise the absence of national sacrifice. The national treasury has been emptied, but private profit is preferred over public interest, while growing political partisanship erodes any sense of national unity. The political class has not served the country well. Or perhaps its constituents have demanded too little of it. In a genuine democracy, after all, the people are responsible for the nation's actions.

At the same time, Americans will defend their liberties. They are ferocious when angered and keen to rise up when thrown on the defensive. And despite the instances of prisoner abuse and torture that have sullied America's honor since 9/11, the American people, on the whole, remain determined to behave virtuously: No cities were leveled after 9/11. Citizens sought to participate in the effort to secure the homeland (but were told to stand aside—and to keep shopping). Most Americans have remained tolerant, although that sentiment is under assault. They are deeply concerned about the country's condition, which they do see as their responsibility to remedy. They are irked by those who blame America for the world's problems and then blame America when it tries to solve them, but in general, Americans continue to believe that their country has an important role to play in the world, and they are eager to play it.

While fighting al Qaeda since 9/11, America has waged a political war with itself. This is nothing new in American life. It may be intrinsic to the nature of our contentious federal republic. But the shadow of 9/11 across America has exacerbated the internal conflicts. Fear may lie at the heart of much of America's response, just as the terrorists intended. But the terrorist attacks have not destroyed America. If anything, they have magnified the extremes within America, from the isolationist impulse to go it alone to the internationalist impulse to remain a beacon of freedom for the world, from the reluctance to engage to the desire to sort things out. In what could be the final legacy of 9/11 and also the most self-defeating consequence of al Qaeda's campaign to extinguish America, the terrorist attacks have compelled America to become an exaggerated version of itself, with its own internal contradictions heightened and intensified. It remains in the hands of the American people to write the next chapter of their history. That is as it should be.

# PART ONE
# Humbled by Hubris

It is as if the United States knew not what to do with victory.

In late 2001, clandestine U.S. intelligence officers and Special Operations advisers, galloping alongside Afghan tribal horsemen and supported by U.S. airpower, ousted the Taliban regime and rousted al Qaeda from its Afghan haven. It was a dazzling campaign hailed as an innovation in modern warfare. But as soon as victory was achieved, it seemed to start slipping away.

The authors in this section offer contrasting explanations for what went wrong. For U.S. Ambassador James Dobbins, who helped negotiate the Bonn agreement forming the new Afghan government in December 2001, subsequent missteps knocked America off its stride. Two major miscalculations were overconfidence in Afghanistan and the overreaction of the Iraq War.

In both Afghanistan and Iraq, U.S. leaders erred in pursuing a tightfisted "low-profile, small-footprint" approach to post-conflict stabilization and reconstruction missions, according to Dobbins. The inadequate resources, especially in Afghanistan, "represented both an exaggerated confidence in the efficacy of high-tech warfare" and "an aversion to the whole concept of nation-building." Left with meager means for building anew, Afghanistan witnessed a resurgence of the Taliban, ironically necessitating far costlier U.S. military investments to fight the growing insurgency.

For Arturo Muñoz, who managed counterterrorism programs at the CIA throughout this period, the paltry investment of U.S. resources in Afghanistan was only one reason for the mission there stalling. Another reason was the *way* in which those resources were applied. "Instead of honoring Afghan terms of peace, utilizing village institutions to maintain security, and training Afghans to do most of their own fighting and rebuilding . . . , the United States and NATO tried to impose Western ways of doing things." The most cavalier misstep of all, says Muñoz, was the U.S. opposition to any reconciliation with the Taliban in early December 2001. "A peace process among the Afghans was being discussed at the time, only to be repudiated by the Americans."

By 2002, the U.S. strategy in Afghanistan shifted to establishing security from the top down by trying to strengthen central government institutions. On the security front, this has meant building the Afghan National Security Forces—consisting of the Afghan National Army, Afghan National Army Air Corps, Afghan National Police, and Afghan Border Police—as the bulwarks against the Taliban and other insurgent groups. On the economic and development fronts, this has meant improving the central government's ability to deliver services to the population.

But "there were few efforts to engage Afghanistan's tribes, subtribes, clans, and other local institutions," laments Seth Jones, who has worked closely with U.S. Special Operations Forces in Afghanistan. From his on-the-ground perspective along the Afghan border with Pakistan, he suggests that America has come full circle in its approach to stabilizing Afghanistan over the past decade. "In recent years, the United States has come a long way, abandoning its overreliance on conventional military forces and returning to clandestine efforts." However, writes Jones, "there is still a long way to go."

The final author in this section, Frederic Wehrey, points to the other key turning point in the battle against al Qaeda: the war in Iraq. An adviser to the U.S.-led Multi-National Force–Iraq in 2008, Wehrey characterizes the 2003 invasion of Iraq and the execution of the Iraq War through 2006 as "an egregious instance of strategic overreach by America." But that is only half of the story, Wehrey explains. From 2006 onward, the strategic overreach in Iraq belonged to a bungling al Qaeda.

The ironies continue. After nearly a decade of grueling U.S. combat operations in Afghanistan and Iraq, the most important turning points of all might have come from Tunisia and Egypt. "Ultimately," observes Wehrey, "the greatest blow to al Qaeda's appeal came not from America's skillful exploitation of the movement's missteps in Iraq, but rather from within the region itself, among the increasingly disaffected youth of authoritarian regimes in North Africa."

Now *that* is a humbling conclusion.

CHAPTER ONE

# The Costs of Overreaction

*James Dobbins*

*U.S. Ambassador James Dobbins, who represented the United States at the Bonn Conference that established the new Afghan government in 2001, directs the International Security and Defense Policy Center within the RAND National Security Research Division.*

The September 11, 2001, attack on the World Trade Center and the Pentagon was unprecedented in the scale of its destruction and the immediacy of its visual impact. Americans had heard or read about other historical disasters, but this was the first to be witnessed by hundreds of millions of citizens as it occurred. The impact on American policy was correspondingly dramatic and long lasting. The immediate impulse was to identify and make pay those who were responsible for the assault. There was an equal determination to make sure that such an attack could never be repeated. But both that impulse and that determination became costlier than originally anticipated.

## Overreactions and Underestimates

It is hardly surprising, given the nature of the 9/11 attacks, that American policymakers overshot in calculating their response. No sooner had Kabul fallen than, in January 2002, President George W. Bush threatened to also attack Iraq, Iran, and North Korea if those regimes did not abandon their nuclear-weapons programs. Indeed, the military planning for the first of those attacks had actually been initiated, at

President Bush's direction, six weeks earlier, while the initial battles to oust the Taliban were still raging.

America's response to 9/11 was also uniquely unilateral in nature. For the previous half-century, all of America's military engagements had been allied campaigns in defense, in the first instance, of others. American troops fought in Korea, Vietnam, and Kuwait; garrisoned Western Europe; and intervened in the Dominican Republic, Lebanon, Grenada, Panama, Somalia, Haiti, Bosnia, and Kosovo to support allies, to liberate oppressed populations, or to maintain global order. As a result, burden-sharing and coalition-building had been essential components of U.S. policy. The U.S. Congress and the American people wanted to be assured that Americans were not alone in shouldering such burdens on behalf of a larger Western and global community. But the American reaction to 9/11 was entirely different. For the first time since Pearl Harbor, the American homeland had been attacked. Americans felt no need to rely on others in the response and scant desire to do so.

These twin impulses of post-9/11 America—to carry the war on terror well beyond those immediately responsible for the 9/11 attacks and to do so largely on its own—marked U.S. policy for the next several years. Washington turned aside most allied offers of military assistance in Afghanistan. It limited international peacekeeping forces there to the Kabul city limits. It invaded Iraq against the advice of several of its most important allies and contrary to the wishes of nearly all the surrounding states that it claimed to be rescuing from potential Iraqi aggression. The Bush administration initially sought to minimize United Nations involvement in postwar Iraq, adopting instead a nominally binational occupation regime alongside the United Kingdom, but one in which London exercised no real authority and was seldom consulted on the major decisions.

Throughout 2002, the Bush administration was far more focused on invading Iraq than on rebuilding Afghanistan. But neither did American military and civilian planners propose to secure Iraq once they had overthrown Saddam Hussein's regime. Consequently, both efforts suffered. In 2002, American troop levels in Afghanistan hovered around 10,000, while the Pentagon's plans for Iraq called for troop levels there to be reduced to 30,000 by September 2003. Afghanistan was the least-resourced American intervention in 50 years, and at the same time, administration officials were assuring the American people that Iraqi reconstruction would be almost entirely self-financing.

These remarkable underestimates represented both an exaggerated confidence in the efficacy of high-tech warfare to cope with low-tech adversaries and an aversion to the whole concept of nation-building. In Afghanistan, the Bush administration not only insisted on confining international peacekeeping forces to Kabul and its immediate environs, ignoring both United Nations and Afghan government pleas to extend the security umbrella to the rest of the country, but also refused to have American troops play any public safety role, insisting that internal security should be an exclusively Afghan responsibility—this in a country which at that point had no army and no police force.

It is not accurate to claim that the Bush administration starved Afghanistan to feed Iraq. Rather, in both cases, the initial American policy was to minimize the American military and economic resources committed to post-conflict stabilization and reconstruction. U.S. Secretary of Defense Donald Rumsfeld justified this "low-profile, small-footprint" approach by arguing that both countries would become self-sufficient more quickly by avoiding excessive dependency on U.S. and other international aid. In fact, the opposite occurred. Faced with the failure of its initial efforts to stabilize both countries, the United States was compelled to vastly increase its commitments of personnel and money. Reinforcing only under the pressure of failure has thus proved a far more expensive approach to nation-building than was the strategy developed in the 1990s of going in heavily, establishing a safe and secure environment quickly, and then drawing down gradually once peace had been established and potential adversaries deterred.

Each of these policies—unilateralism, preemption, and low-profile nation-building—was eventually abandoned, but only after much ground had been lost. By the end of Bush's first term, it was clear that unilateralism had reached its outer limits and was not going to take the United States where it wished to go. It was equally clear that the "invade, overthrow, occupy, and transform" approach to counterterrorism and counterproliferation was too expensive, in blood and treasure, to be applied to Iran, North Korea, or other possible havens of terrorism and nuclear proliferation.

## Bruised but Not Broken

Beginning in 2004, unilateralism gave way to historically more typical efforts to increase burden-sharing. Washington invited, indeed begged, both the United Nations and NATO to play larger roles in Afghanistan and Iraq. International forces that had hitherto been confined to Kabul were reinforced and dispersed throughout the country. In 2005, the Pentagon issued a directive making stability and reconstruction operations a core mission of the American military. That same year, the U.S. State Department established an office of its own for this same purpose, while the White House issued a new presidential directive on the subject. Troop and economic assistance levels climbed in both Iraq and Afghanistan, and the Bush administration embraced nation-building, in all but name, with the fervor of a recent convert.

However, these new resources flowed disproportionately to Iraq, where the security situation was deteriorating alarmingly. In early 2007, President Bush ordered a "surge" of additional troops into Iraq. He assigned the top command there to U.S. General David Petraeus, the author of a new U.S. Army/U.S. Marine Corps manual that put public security at the center of American counterinsurgency doctrine—the polar opposite of Secretary Rumsfeld's "stuff happens" reaction to the initial breakdown of law and order in Iraq four years earlier. By the end of 2007, a sectarian civil war in Iraq had been brought to a close, although terrorist attacks continued. Soon, American troops began slowly to depart. But throughout President Bush's second term, Afghanistan remained an "economy of force operation," a military term that meant, according to the Chairman of the Joint Chiefs of Staff, Admiral Mike Mullen, that "in Iraq, we do what we must, while in Afghanistan, we do what we can."

President Barack Obama thus inherited an improving situation in Iraq but a still-deteriorating one in Afghanistan. Over the objections of some of his advisers, and to the dismay of many of his supporters, Obama chose to replicate President Bush's second-term strategy for Iraq, sending an additional 45,000 American troops to Afghanistan, further increasing economic assistance, and, in 2010, also sending General Petraeus to command the operation.

Bush administration officials had spoken of democracy in Iraq as a priority as early as 2002 and 2003, but President Bush's second-term embrace of nation-building was accompanied by an increasingly fervent emphasis on democracy promotion as the long-term antidote to Islamic extremism. In part, this was a defensive maneuver, democratization then being the last remaining rationale for having invaded Iraq. However, the identification of U.S.-backed democratization efforts with invasions and military occupations (in Iraq and Afghanistan) and with the Israeli occupation of the Palestinian territories rendered Bush's rhetoric rather suspect in Arab eyes. The level of sectarian violence associated with Iraq's liberation was an even stronger deterrent to democracy movements in neighboring countries.

The Bush administration's second-term embrace of democratization thus proved premature but also quite prescient. By the end of the 20th century, every other region of the world except the Arab Middle East had experienced democratic transformation. Almost all of Europe, most of Latin America, and much of Asia and Africa were governed by freely elected regimes. It was thus reasonable to expect that the Middle East would eventually follow suit and to hope that the results of such a transformation would eventually produce a more tranquil region, with less pent-up popular anger and diminished support for extremist movements. The invasion of Iraq probably did more to retard than to advance democracy in the broader region, but in any case, the effect seems to have been transient.

President Obama chose, upon entering office in 2009, to scale back public support for his predecessor's "freedom agenda" rhetoric, while continuing, however, to maintain contact with and to provide quiet assistance to reformist elements of Middle Eastern society. This stance has made it harder for Obama to claim credit for the recent popular revolutions in Tunisia and Egypt, but the lower American profile probably also made it easier for those revolutionaries to garner broad public support in their own societies, freed as they were from the taint of American sponsorship.

One legacy of 9/11 may ultimately be more-or-less democratic regimes in Iraq and Afghanistan, although the costs involved will discourage any attempts to replicate that avenue to democratization elsewhere. President Bush's vision of a democratically transformed Middle East may also be closer to realization now than at any time in history, and this could well prove to be the best antidote to further terrorism emerging from that region. One should not exaggerate the effect of American policy, for better or worse, in bringing about these transformations, however. Democratic values have become widely shared around the globe, and the American example is important in this respect, but American policy is seen by most in the Middle East as at best cynically opportunistic and at worst antidemocratic, a perception that the invasion and botched occupation of Iraq did nothing to correct.

## Related Reading

Allawi, Ali A., *The Occupation of Iraq: Winning the War, Losing the Peace*, New Haven, Conn.: Yale University Press, 2007.

Bremer, L. Paul, III, *My Year in Iraq: The Struggle to Build a Future of Hope*, New York: Simon & Schuster, 2006.

Feith, Douglas J., *War and Decision: Inside the Pentagon at the Dawn of the War on Terrorism*, New York: HarperCollins Publishers, 2008.

Ferguson, Charles H., *No End in Sight: Iraq's Descent into Chaos*, Philadelphia, Pa.: PublicAffairs™, 2008.

Gordon, Michael R., and Gen. Bernard E. Trainor, *Cobra II: The Inside Story of the Invasion and Occupation of Iraq*, New York: Pantheon Books, 2006.

Ricks, Thomas E., *Fiasco: The American Military Adventure in Iraq*, New York: Penguin Press, 2006.

Rudd, Gordon W., *Reconstructing Iraq: Regime Change, Jay Garner, and the ORHA Story*, Lawrence, Kan.: University Press of Kansas, 2011.

Sanchez, Lt. Gen. Ricardo S., *Wiser in Battle: A Soldier's Story*, New York: HarperCollins Publishers, 2008.

Special Inspector General for Iraq Reconstruction (SIGR), *Hard Lessons: The Iraq Reconstruction Experience*, Washington, D.C.: U.S. Government Printing Office, 2009.

Tenet, George, and Bill Harlow, *At the Center of the Storm: My Years at the CIA*, New York: HarperCollins Publishers, 2007.

Woodward, Bob, *Bush at War*, New York: Simon & Schuster, 2002.

———, *Obama's Wars*, New York: Simon & Schuster, 2010.

———, *State of Denial: Bush at War, Part III*, New York: Simon & Schuster, 2006.

———, *The War Within: A Secret White House History 2006–2008*, New York: Simon & Schuster, 2008.

CHAPTER TWO

# A Long-Overdue Adaptation to the Afghan Environment

*Arturo Muñoz*

*Prior to joining RAND as a senior political scientist in 2009, Arturo Muñoz served for 29 years at the Central Intelligence Agency (CIA), where he created and managed counterterrorism, counterinsurgency, and counter-narcotics programs for Latin America, Southwest Asia, the Balkans, the Middle East, and North Africa.*

Small teams of CIA and U.S. Special Operations Forces advisers, working closely with local Afghans, overthrew the Taliban regime in Kabul on November 14, 2001.[1] It was not an invasion. Afghans did all the fighting on the ground, supported by American airpower and high technology, a campaign that has been hailed as an innovation in modern warfare.[2] By making alliances with key Afghan leaders and bringing to bear very intimidating lethal force when needed, these small teams were able to exert political and military influence greatly disproportionate to their number.

As soon as victory was achieved, however, the United States abandoned the advisory model. Instead of honoring Afghan terms of peace,

---

[1] See Gary Schroen, *First In: An Insider's Account of How the CIA Spearheaded the War on Terror*, New York: Presidio Press, 2005; also see Gary Berntsen, *Jawbreaker: The Attack on Bin Laden and Al-Qaeda: A Personal Account by the CIA's Key Field Commander*, New York: Crown Publishers, 2005.

[2] See Henry Crumpton, "Intelligence and War: Afghanistan 2001–2002," in Jennifer Sims and Burton Gerber, eds., *Transforming Intelligence*, Washington, D.C.: Georgetown University Press, 2005, pp. 162–179.

utilizing village institutions to maintain security, and training Afghans to do most of their own fighting and rebuilding from the bottom up, the United States and NATO tried to impose Western ways of doing things from the top down. In doing so, the Westerners received the support of an Afghan ruling elite determined to build a strong central government. However, this approach has often proved to be counterproductive. Even in those cases where counterinsurgency success has been achieved, the crucial concern now is sustainability. The best way to preserve these hard-fought gains is to adapt American military and civilian efforts more closely to Afghan norms.

By advocating inclusion of a bottom-up approach, this essay does not propose abandonment of ongoing top-down efforts to expand the reach of the central government. The two approaches should be combined in a complementary fashion. Local defense forces complement the national army. It is not an either/or proposition. On the contrary, an effective national government is crucial to counterinsurgency success. At the same time, for the national government to achieve its objectives most effectively, it should be responsive to local aspirations and accommodate existing forms of traditional or tribal governance in order to bolster its own authority.

In Afghanistan, an effective aid program—both military and civilian—should not involve a preponderance of Americans or other Westerners assuming leadership roles. Had the advisory role with which the United States began the war been kept in operation, the United States could have secured its vital interests in the region on a more sustainable basis. Afghan traditions, tribal procedures, and methods of conflict resolution should have been incorporated systematically into the U.S. effort from the beginning.

Implementing this approach in practical terms, however, is complex, because the Afghans themselves have different views of what Afghanistan is and should be. Intense rivalries among Afghans do not make it easy to pursue Afghan solutions to Afghan problems. Furthermore, on some issues, traditional Afghan sentiments clash with policies favored by the United States and the European nations that have invested considerable money and personnel in nation-building there. The issue of women's rights is a prime example, as is the role of Islam.

To the great discomfort of the Karzai regime's foreign benefactors, there was widespread popular support in Afghanistan for the proposed death sentence against the Afghan who converted to Christianity (and who was eventually declared insane and deported to Italy). Nonetheless, at a time when the overriding policy mandate is transition, there is little practical alternative but to adapt more effectively to the Afghan environment. This essay argues for an approach in which the West provides respectful assistance to Afghan solutions—what might be called the Afghan-Lead model—in place of the Western-Lead model that has prevailed since late 2001.

## How the Western-Lead Model Has Caused Problems

Prior to the 2009 troop surge, the numbers of U.S. troops in Afghanistan could have been described as insufficient. For years, overall spending was less than what was needed. As the United States began to draw down in Iraq, Afghanistan became a much higher priority and billions of dollars were spent there. But more is not necessarily better in the Afghan context. The *way* we interact with Afghans is critical. The key test of effectiveness is how America is perceived.

In this respect, the heavy concentration of U.S. troops at military bases is instructive. Conventional U.S. military forces began to arrive as soon as December 2001; and on January 3, 2002, the International Security Assistance Force (ISAF) was established in Kabul with 4,500 troops. Bagram and Kandahar military bases, both of which have airstrips accommodating jet bombers, began a process of expansion in 2002 that has continued unabated to the present, with Bagram described today as a military "boom town" with traffic jams. To the Afghans, it looks like a foreign occupation, something they have historically resisted.

One danger of imposing Western predominance on a country such as Afghanistan is the high probability of a nationalist backlash. Since Alexander the Great invaded their land in the third century before Christ, Afghans have resisted foreign soldiers. The more recent experiences of the British and Russians should have given pause to U.S.

policymakers about the idea of fighting a prolonged guerrilla war in Afghanistan with conventional forces. Rather than bringing in American infantrymen to do the fighting, the United States could have built on the successful strategy of small teams and created a systematic program of relying on Special Operations Forces and other advisers to train and supervise local and national Afghan forces.

Another danger of a predominant Western role is the perception, if not the reality, of corruption fueled by large-scale foreign aid. While corruption by Afghan officials has been copiously documented and has become a major factor damaging the legitimacy and effectiveness of the Afghan government, the destabilizing impact of a massive, uncoordinated flow of money into a desperately poor country like Afghanistan is often overlooked. In defending his administration from corruption charges, Afghan President Hamid Karzai reiterates that his government handles directly only about 20 percent of foreign-aid funds, with the remaining 80 percent handled by foreign governments and foreign nongovernmental organizations (NGOs). Directors of Afghan NGOs tend to complain that foreign NGOs receive the bulk of the aid money and siphon off the best local employees with exorbitant salaries. Many Afghans apparently believe that foreign aid organizations spend more money on themselves, their offices, and their SUVs than on the people they are there to help. Afghans do want foreign assistance, not only money but also technology and help in other fields where they see the need for improvement. If this assistance could be offered in a less obtrusive manner consonant with Afghan values, Afghan attitudes toward the foreign presence would probably be much more positive.[3]

In terms of the military, the greatest danger of having U.S. troops play a dominant combat role is that of causing civilian casualties. Public opinion polls show that this is the biggest complaint by Afghans across the board regarding U.S. and NATO forces. As 2010 drew to a close, the controversy of U.S. forces kicking down doors to search

---

[3] See public opinion survey results on attitudes toward foreigners in Helmand and Kandahar provinces in International Council on Security and Development, *Afghanistan: The Relationship Gap*, Brussels, July 2010. As of June 22, 2011: http://www.icosgroup.net/?s=Afghanistan%3A+The+Relationship+Gap

private homes also revived due to expanded Special Operations night raids designed to decimate the Taliban before U.S. troops withdraw. Self-imposed deadlines to show progress have led to a reemphasis on combat operations that unquestionably hurt the Taliban militarily but paradoxically hurt the U.S image as well because of the inevitable civilian casualties.

America risks losing the propaganda war on this issue. The Taliban are responsible for far more civilian deaths than U.S. and NATO forces are, but Afghans are particularly sensitive to the presence of foreign troops, and the killing of Afghans by foreigners generates disproportionate outrage. From the Afghan perspective, even a reduced number of their countrymen being killed by foreigners is unacceptable. Continuing media reports of Americans killing Afghan civilians obscure the fact that the U.S. military takes all sorts of precautions to avoid civilian casualties.

It can be argued that the U.S. troop surge was necessary and is working, citing as evidence considerable progress over the past year in driving the Taliban from targeted communities. When the Marines first went into places like Now Sad and Marjah, the situation seemed hopeless. Today, the Taliban have been pushed out of those towns, district-level government is functioning, economic development is taking place, and Afghan forces are taking over security. A recent public opinion poll among Helmand residents indicates that they are pleased with progress and think their lives will improve.[4] Successes have also been achieved by the U.S. Army in Uruzgan and other provinces. The Taliban are retreating from key population centers they once controlled.

The big concern is over sustainability; that is, how long will the military gains last after U.S. forces leave? This goes to the heart of the debate over a leading versus a supporting U.S. role. American forces now have exerted leadership roles in targeted regions of Afghanistan and have been successful, but it remains to be seen whether the security and economic progress will endure after their departure. Since it

---

[4] See Jeffrey Dressler, *Counterinsurgency in Helmand: Progress and Remaining Challenges*, Washington, D.C.: Institute for the Study of War, Afghanistan Report 8, January 2011. As of May 27, 2011: http://www.understandingwar.org/files/Afghanistan_Report_8_web.pdf

is neither politically nor financially feasible to keep a large American military presence in Afghanistan indefinitely, the U.S. military mission inexorably will devolve to a role of supporting Afghan forces. That being the case, the question arises as to whether it would have been better to play a supporting role from the beginning.

## What an Afghan-Lead Model Could Have Looked Like

First, the United States should have backed Karzai's effort to reconcile with the Taliban in December 2001. A peace process among the Afghans was being discussed at the time, only to be repudiated by the Americans. U.S. reconciliation with the defeated Taliban forces could have prevented them from regrouping and launching an insurgency—and thus could have allowed the United States to maintain the modest military posture with which its intervention began.

Second, much greater emphasis should have been given to training and expanding professional, multiethnic Afghan National Security Forces rather than increasing the reliance on ISAF, which culminated in the current U.S. troop surge. Likewise, a relatively small number of experienced advisers could have accelerated and expanded the training of Afghans as administrators of humanitarian and development projects, rather than bringing in scores of foreign government and NGO personnel to do those jobs.

Third, at the local level, more should have been done to integrate traditional Afghan forms of consensus-based democratic expression (through local *jirgas* and *shuras*)[5] with the centralized, national government apparatus being created in Kabul. And fourth, instead of rejecting the concept of local defense forces, both Afghan and U.S governments should have done more to develop them as force multipliers for efforts to bring security, good governance, and development to rural communities.

---

[5] Historically, a *jirga* is a temporary council established to address specific issues, while a *shura* is a more permanent consultative council. In practice, however, the two terms are often used interchangeably.

Regarding the first element of an Afghan-Lead approach—the peace process with the Taliban—some counterterrorism officials in 2001 wanted to target only al Qaeda. They worried about the danger of getting involved in a prolonged guerrilla war in Afghanistan against native tribesmen. That attitude changed under the "Global War on Terror" agenda, which conflated the Taliban with al Qaeda. Even though, after 9/11, an *ulema* (religious council) of 1,000 Taliban clerics had formally asked that Osama bin Laden leave Afghanistan, reflecting deep divisions within the Taliban over his presence, U.S. policy lumped all Taliban and al Qaeda together, categorizing them as terrorists.[6]

In his capacity as the Pashtun tribal chief who had just been named at Afghan peace talks in Bonn to lead an interim Afghan government, Karzai negotiated a peace plan with the Taliban in early December 2001. In so doing, he followed the *pashtunwali* norm of *nanawatai* (offering sanctuary or reconciliation to defeated enemies). Karzai's peace initiative, launched at the time the Taliban was in disarray and its leaders most receptive to peace talks, was scuttled by U.S. officials.

As the *New York Times* reported on December 7, 2001, "Karzai . . . said that Taliban militants would turn over their arms and ammunition to a council of tribal elders and would be allowed safe passage to their homes. That process, he said, should be completed within a few days." In Islamabad, meanwhile, "Mullah Abdul Salam Zaeef, a Taliban spokesman and former ambassador to Pakistan, announced the surrender agreement had been reached to save civilian lives. 'Tomorrow the Taliban will start surrendering their weapons to Mullah Naqibullah . . . The Taliban were finished as a political force,' said Mullah Zaeef, adding, 'I think we should go home.'" Mullah Zaeef also said that "Mullah Omar would be allowed to live in Kandahar under the protection of Naqibullah," in peace and dignity.[7]

---

[6] Raymond Whitaker, "Council of Clerics Tells bin Laden to Leave the Country," *The Independent*, September 21, 2001. As of May 24, 2011: http://www.independent.co.uk/news/world/asia/council-of-clerics-tells-bin-laden-to-leave-the-country-670126.html

[7] Brian Knowlton, "Rumsfeld Rejects Plan to Allow Mullah Omar 'to Live in Dignity': Taliban Fighters Agree to Surrender Kandahar," *New York Times*, December 7, 2001. As of June 11, 2011: http://www.nytimes.com/2001/12/07/news/07iht-attack_ed3__7.html?pagewanted=1

But U.S. Defense Secretary Donald Rumsfeld quickly abrogated that peace agreement, stating that the United States would not stand for any deal that allowed Taliban leader Mullah Omar "to remain free and 'live in dignity' in the region."[8] Rumsfeld stated his intention to continue military operations in Afghanistan, even though the Taliban had been defeated. Echoing the White House declaration, as stressed in the *New York Times* article, that "those who harbor terrorists need to be brought to justice," Rumsfeld threatened Karzai with loss of support if he persisted in trying to negotiate peace. If any Afghan anti-Taliban leader made a deal with Mullah Omar, Rumsfeld noted pointedly, "our cooperation would take a turn south."[9]

In addition to the obvious political climate in the United States at that time against any peace deal with the Taliban, seen widely as terrorist accomplices, there would have been strong opposition among Afghans themselves regarding any accommodation with Taliban leaders, especially among those who had suffered greatly at their hands. Such accommodation could have provoked a split between the triumphant Northern Alliance and Karzai. Moreover, no one can say with certainty what would have happened had Karzai's peace plan been implemented. Mullah Omar may have surrendered and come to Kandahar only to rebel again for any number of reasons. Some key Taliban commanders may have refused to go along with the plan and kept on fighting on their own, forming the nucleus of a new insurgency. On the other hand, peace could have been established. That possibility cannot be discounted either.[10]

---

[8] Tribal Analysis Center, *"Mizh der beitabora khalq yi": Pashtun Reconciliation Programs*, Williamsburg, Va., July 2008. As of May 27, 2011: http://www.tribalanalysiscenter.com/PDF-TAC/Pashtun%20Reconciliation%20Programs.pdf

[9] Jeremy Armstrong and Andy Lions, "War on Terror: End of Taliban: Give Us Omar; U.S. Demands the Handover of Leader as Kandahar Falls; 'If the Alliance Makes a Deal Our Cooperation Would Turn South'—U.S. Secretary of Defence Donald Rumsfeld Yesterday," *The Mirror* (London, England), December 7, 2001. As of May 27, 2011: http://www.thefreelibrary.com/WAR+ON+TERROR:+END+OF+TALIBAN:+GIVE+US+OMAR%3B+U.S.+demands+the...-a080625065

[10] See Mullah Abdul Salam Zaeef, *My Life with the Taliban*, London: C. Hurst & Company, 2010.

The current U.S. administration is eager to promote peace talks.[11] But times have changed. A decade of war has had a radicalizing effect. The influence of al Qaeda among certain elements of the Taliban is stronger than before, as can be seen in the adoption of terrorist tactics such as suicide bombers, car bombs, attacks on mosques, and the deliberate targeting of civilians, which are not typical of Pashtun warfare. The Taliban are more divided. It will be hard to get them all at the same negotiating table or to make common agreements. Also, a generational gap has emerged in which "neo-Taliban" leaders exemplified by Mullah Abdul Qayum Zakir, who spent six years in Guantanamo, are much more radical than the old leadership. At the end of this long learning curve, America may have learned a lesson that can no longer be implemented.

To achieve the second element of an Afghan-Lead approach—a greater emphasis on training Afghans and building up the Afghan government and military—a close working relationship with American advisers is required. This does not call for large numbers of Americans on the ground. It is better to have fewer people who know what they are doing and are experienced and respected, in the mold of the British political agents of the past in Pakistan. Instead of deploying American combat units to do the fighting, more American trainers should have been brought in.

It has been argued that the *results* of training the Afghan National Security Forces have been disappointing, but this does not take into account that the *training* itself for years was disappointing. Numerous studies and press reports have documented the shortcomings of Afghan National Army and Afghan National Police training. A Government Accountability Office (GAO) study from January 27, 2011, for example, states the following: "High attrition could impact the Afghan National Army's ability to meet its end size goal of 171,600 by October 2011. . . . Efforts to develop Afghan National Army capability have been challenged by difficulties in staffing leadership positions and a shortage of coalition trainers, including a shortfall of approximately 18 percent (275 of 1,495) of the personnel needed to provide instruc-

---

[11] Steve Coll, "U.S.-Taliban Talks," *New Yorker Magazine*, February 28, 2011.

tion at Afghan National Army training facilities. . . . GAO recommends that the Secretary of Defense, in conjunction with international partners, take steps to eliminate the shortage of trainers."[12]

Moreover, when an intense effort is made to provide the best training, as with the training of the Afghan Special Forces, the results are correspondingly positive. An equally intense effort should have been made to develop the Afghan civil service, creating mechanisms for accountability and transparency in handling funds.

Regarding the third element of an Afghan-Lead approach—more reliance on traditional or tribal forms of governance that stress consensus-building—the United States and the West in general have sided with the top-down Kabul elites and have not taken seriously alternatives that could have achieved a more democratic system. A critical challenge in seeking to build on Afghan consensus-building traditions, rather than relying solely on Western-inspired mechanisms, is how to integrate the traditional *jirgas* and *shuras* with national decisionmaking. Efforts to achieve this type of integration were made in convening the 2002 Emergency Loya Jirga and the 2004 Constitutional Loya Jirga. Unfortunately, these grand councils were not used as forums for public debate and consensus-building, but rather to impose behind-closed-doors decisions that had already been made by the regime in power. Two historic opportunities to forge national consensus through a revered traditional mechanism were squandered.

This issue is part of a wider, more fundamental debate regarding the nature of the Afghan state. For centuries, Afghanistan has sought to build a strong central government, and the idea of a "unitary" state along the French model is deeply rooted among Afghan intellectuals, politicians, and bureaucrats. Nonetheless, such an ideal state has not existed in Afghan history. Instead, there has been a constant tension between the efforts of the national government to impose itself and the regional and tribal resistance. The bottom line is that the informal

---

[12] "Afghanistan Security: Afghan Army Growing, but Additional Trainers Needed; Long-Term Costs Not Determined," Washington, D.C.: Government Accountability Office, GAO-11-66, January 27, 2011. As of June 11, 2011: http://www.gao.gov/products/GAO-11-66

*jirgas* and *shuras* in the countryside have never disappeared. They still serve as a potent local forum not only to express opinions but also to rally people for or against the government. Taliban commanders regularly appear before *jirgas* to make their case. American war fighters do the same.

The fourth component of an Afghan-Lead approach would have placed greater reliance on Afghan local defense forces. Although it is true that civilian forces have been badly misused in the past, such forces have worked well in some instances. Afghan kings relied on them throughout history because they were effective. During the Musahiban Dynasty (1929–1978), the Pashtun *arbakai* (traditional village guards) in eastern Afghanistan maintained order successfully, without committing abuses, and protected Afghan control over a disputed border area with Pakistan.

U.S. attempts to develop local defense forces illustrate the difficulty of balancing local and national priorities in Afghanistan. After 2001, as the Taliban turned to guerrilla war, various tribal communities formed *arbakai* on their own to combat them. This is what any good counterinsurgency campaign is looking for and wants to support: local people willing to fight the insurgents. In 2009, to stimulate that process, U.S. General Stanley McChrystal backed the creation of the Local Defense Initiative. However, U.S. Ambassador Karl Eikenberry opposed it, fearing that local forces would inevitably engage in traditional feuding or support warlords. Consequently, the Local Defense Initiative term was abandoned in favor of Village Stability Operations. Based on assessments of the positive impact of Village Stability Operations, U.S. General David Petraeus today strongly supports the program, and a systematic effort organized by the Combined Forces Special Operations Component Command–Afghanistan is under way to expand the number of participating villages on an expedited basis.

However, the Karzai administration has expressed concern over foreigners sponsoring local forces and the potential proliferation of uncontrolled militias. In August 2010, his administration inaugurated its own more-centralized program, the Afghan Local Police, focusing on recruitment and training of local men to become uniformed, salaried policemen controlled by the district or provincial chief of police, under the Afghan Ministry of Interior. To avoid the potential anomaly of two parallel programs, Petraeus signed on to the Afghan Local Police. The mutually agreed-upon compromise calls for the security elements of Village Stability Operations to be absorbed by the Afghan Local Police; that is, members of Village Stability Operations local defense forces eventually will become Afghan Local Police officers. Afghan and U.S. officials currently refer to all local forces as Afghan Local Police, but it remains to be seen whether they will function more as community-based civilian defense forces or as uniformed policemen.

Implementing the four elements of an Afghan-Lead approach could have saved the United States and Afghanistan considerable grief over the past decade. It is still possible for U.S. and Afghan leaders to reconcile with the Taliban, to focus on training Afghan government and military leaders, to integrate tribal with national forms of governance, and to rely more heavily on local defense forces to secure rural communities. But the missteps of the past decade have made each of these elements more difficult to achieve. Nonetheless, they still offer the best hope for bringing long-term peace and stability to Afghanistan.

## Related Reading

Danspeckgruber, Wolfgang, ed., *Building State and Security in Afghanistan*, Princeton, N.J.: Liechtenstein Institute of Self-Determination at Princeton University, 2010.

Jones, Seth G., and Arturo Muñoz, *Afghanistan's Local War: Building Local Defense Forces*, Santa Monica, Calif.: RAND Corporation, MG-1002-MCIA, 2010. As of May 24, 2011:
http://www.rand.org/pubs/monographs/MG1002.html

Koontz, Christopher, ed., *Enduring Voices: Oral Histories of the U.S. Army Experience in Afghanistan 2003–2005*, Washington, D.C.: U.S. Army Center for Military History, 2008.

Meyerle, Jerry, Megan Katt, and Jim Gavrilis, *Counterinsurgency on the Ground in Afghanistan: How Different Units Adapted to Local Conditions*, Washington, D.C.: Center for Naval Analysis, 2010.

Rashid, Ahmed, "How Obama Lost Karzai: The Road Out of Afghanistan Runs Through Two Presidents Who Just Don't Get Along," *Foreign Policy*, March/April 2011.

CHAPTER THREE

# Lessons from the Tribal Areas

*Seth G. Jones*

*A senior political scientist at the RAND Corporation, Seth G. Jones recently served as the representative for the commander, U.S. Special Operations Command, to the U.S. Assistant Secretary of Defense for Special Operations. Prior to that assignment, Jones served as a plans officer and adviser to the commanding general of U.S. Special Operations Forces in Afghanistan. He is the author of* In the Graveyard of Empires: America's War in Afghanistan *(2009).*

I stepped off a Blackhawk helicopter in early 2011 in Bermel, Afghanistan, several miles from the border with Pakistan's North and South Waziristan Agencies. The jagged limestone mountain ranges, cavernous gorges, sparse population, and parched landscape make it inhospitable terrain and an ideal place for a terrorist sanctuary. As I had witnessed on previous trips, the area has a haunting, isolated feel. "The quietness of the place is uncanny," wrote T. E. Lawrence, better known as Lawrence of Arabia, during a short visit in the 1920s, "no birds or beasts except a jackal concert for five minutes about ten p.m." Since September 11, 2001, the Afghanistan-Pakistan border has been the epicenter of al Qaeda's global sanctuary and the focus of U.S. counterterrorist operations.

I was in Afghanistan to assess the current state of al Qaeda and the effectiveness of U.S. operations in the region. Nearly ten years after September 11, 2001, two controversial questions persist: How serious is the threat to the United States from al Qaeda and its affiliated groups in the Afghanistan-Pakistan border region? And what lessons, if any, has the United States learned in its pursuit of al Qaeda?

After my trek through this region, the short answers seemed clear, though perhaps anodyne. Central al Qaeda, based in the Afghanistan-Pakistan border region, continues to pose a threat, though a weakened one, to the U.S. homeland. So do a range of affiliated groups. Yet there has been a steep learning curve over the past decade in countering al Qaeda and its allies in the region. Perhaps the most significant lesson has been the futility of imposing stability *only* from the center in a region where power, especially in rural areas, remains local and where individuals identify themselves by their tribe, sub-tribe, clan, or community. Rather than banking on stability entirely from the top down, a better strategy has involved developing complementary bottom-up efforts that co-opt local communities against the Taliban, al Qaeda, and their allies.

The short answers have large implications. It has taken a long time for America to settle into a strategy that appears to be working once again—nearly a decade after the Taliban were swiftly ousted from power by small U.S. military and intelligence teams fighting alongside indigenous forces. In the interim, thousands of lives have been lost. The struggle against al Qaeda and its sympathizers is often characterized as the "long war," but it has likely become longer than necessary because of a naïveté about the region and a failure to learn lessons from the past.

## Debating the Threat

Since September 11, 2001, there has been growing skepticism about the importance of Afghanistan and Pakistan for U.S. national security. Some have argued that al Qaeda has a nearly endless supply of sanctuaries in weak states, such as Yemen, Somalia, Djibouti, Sudan, and even Iraq. "Many of these countries," notes Stephen Biddle from the Council on Foreign Relations, "could offer al-Qaeda better havens than Afghanistan ever did."

While this argument seems reasonable on the surface, the evidence suggests that al Qaeda leaders retain an unparalleled relationship with local networks in the Afghanistan-Pakistan frontier. Indeed, Ayman al-Zawahiri and others have a 30-year, unique history of trust and collaboration with the Pashtun militant networks located in North and South Waziristan, in Bajaur Agency (also in Pakistan), and in northeastern Afghanistan. These relationships are deeper and stronger than the comparatively nascent, tenuous, and fluid relationships that al Qaeda has established with al Shabaab militants in Somalia and local tribes in Yemen. In addition, moving from Afghanistan and Pakistan to other areas makes al Qaeda operatives increasingly vulnerable to intelligence-collection and targeting.

Al Qaeda has become embedded in multiple networks that operate on both sides of the Afghanistan-Pakistan border. Key groups straddling the border include the Tehrik-e-Taliban Pakistan, Haqqani Network, Lashkar-e Tayyiba, and Afghan Taliban. Al Qaeda has established a foothold with some of the Mahsud sub-tribes of the Tehrik-e-Taliban Pakistan, the Zadran sub-tribes of the Haqqani Network, and the Salarzai and other sub-tribes of militants in northeastern Afghanistan. It also retains a strategic relationship with senior officials from the Afghan Taliban, including Taliban leader Mullah Omar and military chief Mullah Zakir. The secret to al Qaeda's staying power, it turns out, is its success in cultivating supportive networks in an area generally inhospitable to outsiders. "Except at the times of sowing and of harvest, a continual state of feud and strife prevails throughout the land," Winston Churchill wrote of the region in his epic tome, *The Story of the Malakand Field Force*. "Every tribesman has a blood feud with his neighbour. Every man's hand is against the other, and all against the stranger."

Al Qaeda provides several types of assistance to local groups in return for sanctuary. One is military coordination. It has helped establish *shuras* (councils) to coordinate strategic priorities, operational campaigns, and tactics against U.S. and coalition forces. Al Qaeda operatives have been involved in launching suicide attacks, emplacing improvised explosive devices, and helping conduct ambushes and raids. Al Qaeda helps run training camps for militants, which cover the recruitment and preparation of suicide bombers, intelligence, media and propaganda efforts, bomb-making, and religious indoctrination. It provides some financial aid to militant groups, though it appears to be a small percentage of their total aid. Al Qaeda has also cooperated with Afghan and Pakistani militant groups to improve and coordinate propaganda efforts through the use of DVDs, CDs, jihadi websites, and other media.

In addition to the long-standing relationships in this region, the geography itself presents a formidable challenge. Though I was technically standing on Afghan soil, peering across the mountains of South and North Waziristan atop a watchtower constructed of corrugated iron and plywood, the border is a ruse. Some pundits have argued that al Qaeda operatives primarily reside in Pakistan, not Afghanistan. But the 1,519-mile border, drawn up in 1893 by Sir Henry Mortimer Durand, the British Foreign Secretary of India, is largely irrelevant. Locals regularly cross the border to trade, pray at mosques, visit relatives, and—in some cases—target U.S. and coalition forces. Indeed, al Qaeda migration patterns have shown frequent movement in both directions ever since the anti-Soviet jihad. Osama bin Laden established al Qaeda in Peshawar, Pakistan, in 1988, and he and other Arab fighters crossed the border into Afghanistan regularly to fight Soviet forces and support the mujahedeen. When bin Laden returned to the area in 1996 from Sudan, he settled near Jalalabad in eastern Afghanistan and later moved south to Kandahar province. After the overthrow of the Taliban regime, most of the al Qaeda leadership moved back to Pakistan, though some settled in neighboring Iran.

The patterns endure. "Even today," one U.S. Special Operations soldier remarked, as we flew in a Blackhawk helicopter from Bermel north along the border to Asadabad, "some al Qaeda operatives have pushed back across the border into Afghan provinces such as Konar and Nangarhar. They look for opportunities where there are supportive local communities and few coalition forces." Asadabad is in Konar province, where the rugged, snow-capped Hindu Kush mountains, narrow valleys, and swelling rivers along both sides of the border provide another ideal sanctuary for al Qaeda and other militants. In

January 2006, a U.S. drone tried—and failed—to target Ayman al-Zawahiri across the border from Konar in Pakistan's Bajaur Agency.

Still, skeptics downplay the importance of the Afghanistan-Pakistan front. Some, such as former CIA operative Marc Sagemen, contend that informal, homegrown networks inspired by al Qaeda have become the most pernicious threat to the United States. Ayman al-Zawahiri and al Qaeda Central have become extraneous, according to this argument. Impressionable young Muslims can radicalize through the Internet or interactions with local extremist networks. They don't need a headquarters. The threat to the United States, therefore, comes largely from a "leaderless jihad" in Europe, Asia, the Middle East, and North America, rather than from a relationship with central al Qaeda located in Afghanistan and Pakistan.

But there is sparse evidence to support this argument. Many of the recent terrorist plots against the United States and its allies have been connected to al Qaeda and its affiliates in the Afghanistan-Pakistan border region. A few others have been tied to such areas as Yemen. Few serious plots have come from purely homegrown terrorists.

In December 2009, five Americans from Alexandria, Virginia, were arrested in Pakistan and charged with plotting terrorist attacks. In February 2010, Najibullah Zazi pleaded guilty in U.S. District Court to "conspiracy to use weapons of mass destruction" and "providing material support for a foreign terrorist organization" based in Pakistan. According to U.S. government documents, Zazi's travels to Pakistan and his contacts with individuals there were pivotal in helping him build an improvised explosive device containing triacetone triperoxide, the same explosive used in the 2005 London subway bombings. Several al Qaeda operatives, including Saleh al-Somali and Adnan Gulshair el Shukrijumah, were involved in the plot. In May 2010, Faisal Shahzad attempted to detonate a car bomb in Times Square in New York City after being trained by Tehrik-e-Taliban Pakistan bomb-makers in Waziristan, not far from where I stood in Bermel.

The same is true for many of America's staunchest allies. Jonathan Evans, the Director General of MI5, the United Kingdom's domestic intelligence agency, acknowledged that at least half of his country's most serious plots continue to be linked to "al Qaeda in the tribal areas of Pakistan, where al Qaeda senior leadership is still based." Over the past decade, there has been a laundry list of plots and attacks in the United Kingdom, Germany, Spain, the Netherlands, France, India, and other countries, all with links to al Qaeda and other terrorist groups in the Afghanistan-Pakistan border region.

## Countering al Qaeda and Its Allies

In 2001, indigenous Afghan groups, supported by fewer than 100 CIA and U.S. Special Operations personnel and punishing U.S. airpower, toppled the Taliban regime and unhinged al Qaeda from Afghanistan. By 2002, however, the U.S. strategy shifted to establishing stability only from the top down by trying to strengthen central government institutions. On the economic and development fronts, this strategy involved improving the central government's ability to deliver services to the population. On the security front, it focused on building Afghan National Police and Afghan National Army forces as the bulwarks against Taliban and other insurgent groups. Yet this strategy soon ran into problems, especially with the Afghan police. In 2005, the U.S. Department of Defense took over the police training program from the U.S. Department of State and from Germany because the Afghan police force had essentially collapsed.

Meanwhile, there were few efforts to engage Afghanistan's tribes, sub-tribes, clans, or other local institutions. Masses of rural Afghans today still reject having a strong central government actively meddling in their affairs. In southern and eastern Afghanistan, which are dominated by Pashtuns, many consider the central government a foreign entity. "My allegiance is to my family first," one tribal elder from Kandahar told me. "Then to my village, sub-tribe, and tribe," he continued, noting that the government played no meaningful role in his daily life.

Tribal, religious, and other local leaders in Afghanistan best understand their community needs, but they are often underresourced or intimidated by Taliban, al Qaeda, and other insurgents. This is where the Afghan and U.S. governments have recently helped. A key starting point is security and justice. In some areas, local tribes and villages have already tried to resist the Taliban but have been heavily outmatched. The solution should be obvious: The local tribes and villages should be strongly supported.

The initial efforts have included establishing village-level "community watch" programs rooted in the *shura*, the legitimate governing institution in Pashtun areas. In some places, *shuras* are composed primarily of tribal leaders, who adjudicate disputes and mete out justice. In other places, *shuras* include religious and other figures. Finding ways for organizations such as the Afghan army to support these village-level defense programs—for example, by helping them develop quick-reaction forces that can respond when the villages come under attack—would give the people a reason to ally with the central government.

By 2010, Afghan and U.S. policymakers seemed to have a growing appreciation for the local nature of politics in Afghanistan, establishing the Afghan Local Police program to identify grassroots resistance and to help train and equip Afghan communities to defend themselves against insurgents. The United States also increased its covert efforts against al Qaeda, improving its intelligence-collection capabilities and nearly tripling the number of drone strikes in Pakistan from 2009 levels. Recognizing the importance of al Qaeda's local hosts, the United States stepped up efforts to recruit fighters among rival subtribes and clans in the border areas.

The results have been encouraging. In Afghanistan, intelligence and Special Operations activities have disrupted al Qaeda, making it less cohesive and more decentralized among its surviving assortment of foreign fighters. Al Qaeda has retained a minimal presence in Afghanistan, with perhaps fewer than 100 full-time native-born fighters at any one time. The number is larger if one counts al Qaeda–affiliated networks of foreign fighters operating in Afghanistan. In Pakistan, several senior-level al Qaeda officials have been killed, including Osama bin Laden and chief financial officer Shaykh Sa'aid al-Masri. This has left perhaps fewer than 300 native Pakistani al Qaeda members in the country, although there are larger numbers of foreign fighters and affiliated organizations. In late 2010, Ayman al-Zawahiri ordered al Qaeda operatives to disperse into small groups in Afghanistan and Pakistan, away from the tribal areas, and to cease most activities for up to one year to ensure the organization's survival.

What does this progress mean? For starters, the number of al Qaeda operatives in Afghanistan and Pakistan combined has shrunk from the level in 2001, when it was probably over 1,000. More importantly, U.S. efforts have disrupted al Qaeda's command and control, communications, morale, freedom of movement, and fund-raising. Central al Qaeda is a notably weaker organization. While not defeated, al Qaeda has failed to pull off a successful terrorist attack in the West since July 2005, when Mohammad Sidique Khan and three other suicide bombers staged devastating attacks in London, killing 56 people and injuring more than 700 others, many of whom became amputees. Al Qaeda has tried—but repeatedly failed—to conduct a follow-on attack in the United States. In addition, the death of senior leaders has forced al Qaeda to become increasingly reliant on couriers, has hampered communication because of operational security concerns, has delayed the planning cycle for operations, and has exposed operations to interdiction.

## A Long War

Wandering along the border in Bermel, I remained upbeat in an otherwise bleak environment. The landscape is strangely reminiscent of those Frederick Remington or C. M. Russell paintings of the American West. Gritty layers of dust sap the life from a parched landscape. With the exception of a few apple orchards, there is little agricultural activity, because the soil is too infertile. Several dirt roads snake through the area. Virtually no roads are paved. Yet in this austere, unwelcoming environment, central al Qaeda has been disrupted. In recent years, the United States has come a long way, abandoning its overreliance on conventional military forces and returning to clandestine efforts by U.S. intelligence agencies and Special Operations forces.

In addition, support for al Qaeda has begun to dwindle. Public perceptions of al Qaeda throughout the Muslim world have plummeted. According to a 2010 public opinion poll published by the

New America Foundation, more than three-quarters of the residents in Pakistan's Federally Administered Tribal Areas oppose the presence of al Qaeda. A poll conducted by the Pew Research Center indicated that between 2001 and 2011, positive views of al Qaeda significantly declined across the Middle East and Asia, including in Indonesia, Jordan, Pakistan, Turkey, Egypt, and Lebanon. There was widespread opposition to al Qaeda's ideology and tactics, especially the practice of killing civilians, even among conservative Islamic groups. Public opposition to al Qaeda, especially from legitimate Muslim religious leaders, needs to be better encouraged, supported, and publicized.

There is still a long way to go. As our Blackhawk helicopter departed Bermel, the sun was slowly setting, its last rays spilling over the silhouetted hills to the west. "We are getting closer," remarked one U.S. Special Operations soldier sitting next to me, sounding hopeful. "But it will still be a long war." He was right, of course, even with the death of Osama bin Laden. As Winston Churchill observed more than a century ago during the British struggles in the Northwest Frontier, time here is measured in decades, not months or years. It is a concept that does not always come easily to Americans.

## Related Reading

Bergen, Peter, *The Longest War: The Enduring Conflict Between America and Al-Qaeda,* New York: Free Press, 2011.

Berntsen, Gary, *Jawbreaker: The Attack on Bin Laden and Al-Qaeda: A Personal Account by the CIA's Key Field Commander*, New York: Broadway, 2006.

Churchill, Winston, *The Story of the Malakand Field Force: An Episode of Frontier War*, London: Leo Cooper, 1989.

Coll, Steve, Ghost Wars: *The Secret History of the CIA, Afghanistan, and bin Laden, from the Soviet Invasion to September 10, 2001*, New York: Penguin, 2004.

Dobbins, James, *After the Taliban: Nation-Building in Afghanistan*, Washington, D.C.: Potomac Books, 2008.

Jones, Seth G., *In the Graveyard of Empires: America's War in Afghanistan*, New York: W.W. Norton, 2009.

Jones, Seth G., and Arturo Muñoz, *Afghanistan's Local War: Building Local Defense Forces*, Santa Monica, Calif.: RAND Corporation, MG-1002-MCIA, 2010. As of May 24, 2011:
http://www.rand.org/pubs/monographs/MG1002.html

Neumann, Ronald, *The Other War: Winning and Losing in Afghanistan*, Washington, D.C.: Potomac Books, 2009.

Rashid, Ahmed, *Descent into Chaos: The U.S. and the Disaster in Pakistan, Afghanistan, and Central Asia*, New York: Penguin, 2009.

Schroen, Gary, *First In: An Insider's Account of How the CIA Spearheaded the War on Terror in Afghanistan*, New York: Presidio Press, 2007.

CHAPTER FOUR

# The Iraq War: Strategic Overreach by America—and Also al Qaeda

*Frederic Wehrey*

*A RAND senior policy analyst, Frederic Wehrey served as an adviser to the Multi-National Force–Iraq in 2008. He has also served as an active-duty and reserve U.S. Air Force officer in Turkey, Iraq, Libya, and elsewhere in the Middle East. He is the co-author of* The Iraq Effect: The Middle East After the Iraq War *(2010).*

Common wisdom suggests that the Iraq War was an egregious instance of strategic overreach by America—a bridge too far in the struggle against terrorism, which had uniformly deleterious effects. The invasion and the botched occupation, the narrative goes, injected new life into al Qaeda at precisely the point when its safe havens were under attack and its strategy of attacking the "far enemy" was under fierce criticism from even diehard supporters. Yet this assessment, while accurately describing the period of 2003 to 2006, is only half of the picture. From 2006 to the close of the U.S. drawdown in Iraq, al Qaeda's strategic missteps, along with a skillful shift in U.S. strategy that capitalized on growing tribal resentment, left the movement increasingly exhausted, isolated, and rife with fissures.

In the final reckoning, the outcome of the Iraq War hardly matched the optimistic expectations of its planners or the post-facto justifications of its supporters. The toppling of Saddam Hussein did not create a wave of liberalization in the Middle East that "drained the swamp" of radicalism. Nor did it provide an opportunity for the United States to tie down al Qaeda and decimate its numbers en masse.

But the war also did not fulfill the apocalyptic fears of its detractors. Instead, the eight-year conflict brought to the surface long-standing and potentially fatal tensions inherent in al Qaeda: between national and transnational agendas; between tactical expediency and strategic patience; between the ambitions of the local *amir*s, or chieftains, and the grand designs of al Qaeda's leadership. Ultimately, the movement's fratricidal conduct in Iraq created "too many enemies," inflicting irreparable damage on its popular standing.

For its part, the United States skillfully exploited this alienation, both in Iraq and elsewhere in the region, by encouraging local actors to take the lead against al Qaeda, whether those actors were grassroots Awakening Councils in Iraq or allied Arab governments. A key turning point, however, came not from any U.S. efforts to recover the momentum in Iraq, but from developments in North Africa. The early 2011 revolts in Tunisia and Egypt effectively overturned al Qaeda's long-standing arguments about the necessity of armed jihad as a tactic to overthrow authoritarian regimes and the United States' immutable commitment to propping up those regimes. The revolts also showed that the region has its own antibodies to al Qaeda, and the United States would be wise to let those play out on their own.

## The Narrative Godsend to al Qaeda

Among the initial boosts the 2003 invasion gave to al Qaeda, none is as significant as the public relations coup it handed the movement. The invasion by a Western force of an Islamic, oil-rich land that had once served as the seat of the mighty Abbasid caliphate seemed a powerful vindication of al Qaeda's argument that the West was irrevocably bent on subjugating Muslims. The ensuing conflict enabled al Qaeda to refashion the narrative of its struggle into a defensive jihad—a war of liberation in an Islamic land, compelling able believers outside Iraq's borders to take up arms. The chaos that ensued after the poorly planned occupation allowed al Qaeda to expand its recruitment base beyond its traditional home in the Arabian Peninsula and Egypt, to the Levant and the Maghreb. The war also radicalized a new cadre of

Iraqi salafi-jihadists, or violent Islamic fundamentalists, where few had existed before.

Aiding this recruitment, the war created a broader backlash against American presence in the Middle East and, more ominously, against the cachet of democratization itself. The conduct of America's counterinsurgency campaign was in many ways a powerful boost to al Qaeda's rampant anti-Americanism, with the indelible images of Abu Ghraib creating fertile ground for the recruitment of extremists. Governments around the region found a new impetus to delay or defer their steps toward reform and liberalization by pointing to the civil war in Iraq, which they argued had been sparked by the forcible and premature imposition of democracy. In many public opinion surveys, democratization during the Iraq War assumed a negative connotation because of its affiliation with the American misadventure there, the difficulties of building a cohesive government in Iraq, and the near universal unpopularity of the Bush administration.

## The University of Jihad?

The Iraq War proved to be a new laboratory for guerrilla warfare, clandestine organization, and the development of lethal tactics against conventional armies—tactics such as improvised explosive devices and suicide bombings. In the words of one jihadi ideologue, "If [the 1980s war in] Afghanistan was the school of jihad, then Iraq is the university." Yet widespread fears of a so-called jihadi "bleed out"—the exodus of hardened, battle-trained "graduates" to other theaters—did not materialize as expected. In many cases, foreign volunteers in Iraq did not acquire significant skills that were transferrable to other battlefronts. This was partly because of the arduous and unique conditions of urban warfare in Iraq, but also because jihadi commanders in Iraq tended to use poorly trained but ideologically committed foreigners as cannon fodder.

The accounts of returning veterans suggest that far from being the noble experience of jihad they had expected, their service in Iraq was confused, demoralizing, and ultimately unfulfilling. Many returned disillusioned to their countries of origin. Governments throughout the region proved effective in exploiting this disillusionment through reindoctrination and resettlement programs. Saudi Arabia in particular was able to use the experiences of returning veterans to overturn the narrative of jihadi volunteerism that had existed in the kingdom from the time of the Soviet invasion of Afghanistan through the wars in Bosnia and Chechnya. Other reindoctrination programs were present in Yemen, Morocco, Jordan, and Libya.

Perhaps the most significant example of jihadi veterans from Iraq actually applying their expertise to their countries of origin was Lebanon. Here, returning jihadists led by a Palestinian, Shakr al-Absi, a onetime lieutenant of Abu Musab al-Zarqawi in Iraq, proved instrumental in spreading combat expertise to the country's Palestinian refugee camps, which had long existed as no-go zones for the Lebanese government. The influx of these fighters ultimately led to a showdown with the Lebanese government at the Nahr al-Barid refugee camp in 2007, severely testing the capabilities of the Lebanese army and the very unity of the Lebanese state.

## The (Mis)management of Savagery

In many ways, the event that tipped the scales against al Qaeda's campaign in Iraq was the emergence of its most prominent battlefield commander, the aforementioned Abu Musab al-Zarqawi. The Jordanian jihadist, hardened by years of prison and battle-tested in Afghanistan, arrived in northern Iraq before the invasion. The various mutations of his insurgent organization quickly established a reputation as the most fearless and brutal of the insurgent groups. His ferocious tactics of beheadings, suicide bombings against civilians, and targeting Shiite Muslims were initially a recruiting windfall, particularly after the United States lavished excessive attention on him, granting him a degree of notoriety he would otherwise not have enjoyed. On the surface, al-Zarqawi's strategy appeared to conform neatly to that envisioned by the notorious al Qaeda theorist Abu Bakr Naji in his lengthy blueprint for global jihad, *The Management of Savagery*. Sow enough *fitna* (chaos) and bloodshed, Naji advised, and the civilian population will welcome jihadists with open arms as liberators and providers of security.

In fact, the opposite occurred. Al-Zarqawi's increasingly indiscriminate violence ultimately proved to be his undoing. This was especially so after he tried to extend the war beyond the borders of Iraq. His bombings of hotels in Amman, Jordan, in 2005, most of whose victims were Jordanian civilians, spawned a sharp outcry throughout the Arab world and among many jihadi ideologues. This atrocity was followed by a similar attack on civilians by al Qaeda's affiliates in Saudi Arabia, which provoked similar outrage. The net result was that populations that had previously cheered al Qaeda from afar turned against it when confronted with its violence at home.

After al-Zarqawi's death from a U.S. air strike in 2006, al Qaeda's campaign continued to deplete its public support inside Iraq and beyond. In the western Iraqi province of al Anbar, where the group had once enjoyed great maneuverability due to the marginalization of the area's Sunni tribes from the newly Shiite-led government, al Qaeda's attempt to govern the area was met with growing resistance starting in 2008. The tribes chafed under al Qaeda's draconian social mores, extortion, and, most importantly, attempts to marry local women. The tribal backlash led to the formation of the so-called "awakening councils" and "sons of Iraq"—Sunni Iraqi militias that the U.S. forces adroitly exploited to drive al Qaeda from the area and diminish its logistical and funding networks.

## Lessons Learned by al Qaeda . . .

Al-Zarqawi's tactics and the broader experience of al Qaeda in Iraq stirred unprecedented debate among jihadi strategists and ideologues, creating disarray and strategic gridlock. This was most evident in the sharp rebukes to al-Zarqawi by Ayman al-Zawahiri and, importantly, the Jordanian jihadist's former spiritual mentor, Abu Muhammad al-Maqdisi. Both figures accused al-Zarqawi of needlessly antagonizing broad segments of not only Iraqi society but also the broader Muslim world. Al-Zawahiri, in particular, appeared to demonstrate an acute awareness that the war was increasingly being fought in the realm of media and public opinion—a center of gravity that al-Zarqawi seemed to ignore.

Senior and respected clerics in Saudi Arabia, Jordan, Egypt, and Europe joined the chorus of criticism. Many of these figures had at one time supported the legitimacy of defensive jihad against an occupying force, but they now turned their attention to a thorough deconstruction of al-Zarqawi's tactics and al Qaeda's broader strategy in Iraq. Marshaling broad-ranging theological support from the Koran, the *hadith* (narrations of the life of Muhammad), and revered scholars like Ibn Taymiya, these clerics undermined the legitimacy of suicide bombings and of attacking civilians and fellow Muslims. These efforts among religious clerics, often sponsored by Arab governments, were amplified by the near-simultaneous appearance of a number of recantations by former jihadi commanders and ideologues.

## . . . and the United States

With regard to the United States, the Iraq War proved indeed to be a regrettable diversion of resources, expertise, and manpower away from what had been al Qaeda's central haven, Afghanistan. The neglect of the Afghan front enabled the Taliban insurgency to recover from the setbacks it had suffered in the immediate aftermath of the 2001 U.S. invasion of the country. That said, the Iraq campaign did impart to U.S. commanders a degree of strategic and tactical learning that has had application not only in Afghanistan but also in potential theaters beyond, where terrorists may seek to capitalize on long-standing tribal grievances and nascent insurgencies.

Specifically, the war demonstrated the imperative of securing the population and providing order as a pillar of counterinsurgency. The war also highlighted the importance of differentiating between more-nationalist-oriented insurgents and the less-reconcilable foreign jihadists. Similarly, the conflict underscored the necessity of capitalizing on locally recruited militias and tribal legitimacy as a bulwark against foreign al Qaeda elements.

## The Iraq War Versus the Arab Uprisings

The drawdown of U.S. forces in Iraq and the election of Barack Obama as U.S. president created a new context for a diminution of al Qaeda's narrative. Obama's well-received speech in Cairo in 2009 was generally seen as a promise to shift from trying to impose American will on the region to accepting the desirability and inevitability of change from within. From al Qaeda's perspective, the speech and the planned withdrawal from Iraq were hailed as admission of defeat and validation of armed jihad. Yet this triumphalism proved short-lived. Ultimately, the greatest blow to al Qaeda's appeal came not from America's skillful exploitation of the movement's missteps in Iraq, but rather from within the region itself, among the increasingly disaffected youth living under authoritarian regimes in North Africa.

Tunisia's revolt in early 2011 and the subsequent ousting of Egypt's Hosni Mubarak dealt a significant blow to al Qaeda's narrative. The popular toppling of two long-despised rulers and, most importantly, the U.S. support for the aspirations of the Tunisian and Egyptian people undermined two of al Qaeda's most potent arguments. The first was that authoritarian rulers in the Middle East would never be ejected by peaceful means and that the only viable strategy was armed jihad against either the regimes themselves or their principal patron, the United States. The second was that the United States was deeply and immutably committed to propping up these despots against the wishes of their people. Despite the belated attempts at "spin" by al Qaeda's propagandists—including the late Osama bin Laden himself—the revolts have shaken the jihadist worldview and left the movement grasping for new direction.

## Related Reading

al-Baghdadi, Abu Umar, emir of the Islamic State of Iraq, "Al-Qa'ida: al-'Iraq Jami'at al-Irhab" (al-Qa'ida: Iraq Is University of Terrorism), April 17, 2008. As of May 24, 2011:
http://www.middle-east-online.com/?id=47152

Bergen, Peter, and Paul Cruickshank, "The Iraq Effect," *Mother Jones*, March/April 2007.

Byman, Daniel L., and Kenneth M. Pollack, "Iraq's Long Term Impact on Jihadist Terrorism," *Annals of the American Academy of Political and Social Science*, July 2008, Vol. 618, No. 1, pp. 55–68.

*Declassified Key Judgments of the National Intelligence Estimate "Trends in Global Terrorism: Implications for the United States,"* April 2006. As of May 24, 2011:
http://hosted.ap.org/specials/interactives/wdc/documents/terrorism/keyjudgments_092606.pdf

Fishman, Brian, "After Zarqawi: The Dilemmas of Future of al Qaeda in Iraq," *The Washington Quarterly*, Autumn 2006.

———, ed., *Bombers, Bank Accounts and Bleedout: Al-Qaida's Road In and Out of Iraq*, West Point, N.Y.: Combating Terrorism Center at West Point, July 2008. As of May 24, 2011:
http://www.ctc.usma.edu/harmony/pdf/Sinjar_2_July_23.pdf

Hafez, Muhammad, "Lessons from the Arab-Afghans," *Studies in Conflict and Terrorism*, 2009, Vol. 32, pp. 73–94.

Hiltermann, Joost, "Iraq and the New Sectarianism in the Middle East," synopsis of a presentation at the Massachusetts Institute of Technology, November 12, 2006. As of July 3, 2011:
http://www.crisisgroup.org/en/publication-type/speeches/2006/iraq-and-the-new-sectarianism-in-the-middle-east.aspx

Naji, Abu Bakr, *Idarat al-Tawahhush (The Management of Savagery)*, translated by William F. McCants, under sponsorship by the John M. Olin Institute for Strategic Studies, Harvard University, May 23, 2006. As of May 24, 2011:
http://www.wcfia.harvard.edu/olin/images/Management%20of%20Savagery%20-%2005-23-2006.pdf

Wehrey, Frederic M., Dalia Dassa Kaye, Jeffrey Martini, Jessica Watkins, and Robert Guffey, *The Iraq Effect: The Middle East After the Iraq War*, Santa Monica, Calif.: RAND Corporation, MG-892-AF, 2010. As of May 24, 2011:
http://www.rand.org/pubs/monographs/MG892.html

# PART TWO
# Hopeful amid Extreme Ideologies and Intense Fears

It is remarkable how much of the war against violent jihadist terrorism is waged on an intangible battleground. Fueled by ideology, the terrorists take aim at psychology. Meanwhile, one of the most formidable counterattacks could come from theology.

"At its core the challenge of radical Islamism is ideological," writes Angel Rabasa. "To prevail against the challenge of radical Islamism, therefore, it is necessary to address comprehensively the ideological challenge that it presents—the content or substance of the ideology, its ideological agents, networks, and means of dissemination—and to devise effective countermeasures for each of these components." Rabasa argues for giving moderate Muslim leaders the tools they need to contest the ideological ground within their communities.

Key Muslim leaders have begun to do just that, according to Eric Larson, who offers a litany of antiterrorist invocations emanating from respected Islamic theologians. This contest "over the true nature of Islam," observes Larson, "has aptly been described both as a civil war within Islam itself and as being analogous to the West's own century-long Reformation and Counter-Reformation of the 16th and 17th centuries." In this "intra-Muslim ideological battle," he says, "al Qaeda appears to be vastly outgunned."

Terrorism is a strategy for the weak, for those whose chief weapon is a spurious claim to divine retribution. It is a strategy limited only by the fantasies of its perpetrators—and by the fears of its victims. It is also a strategy suited to an age of global communication short on information.

"We live at the edge of doom" is how Brian Michael Jenkins describes the popular mood. The most dreaded fate of all is an impending nuclear doom of our own imagining. "The threat of nuclear terror floats far above the world of known facts." The fact that terrorist possession of a nuclear weapon remains purely hypothetical becomes lost among the "reality" of its potential effects, particularly when the news media veer off "into the realm of shock and entertainment."

The "avid consumption" of nuclear terror speaks volumes about American society, says Jenkins. But he detects a source of hope even within that terror. "The heightened fears, in particular, may be stimulating nations worldwide, and factions within them, to become more unified in demanding and supporting stringent measures to control nuclear weapons and related materials."

The battlegrounds of the intangible offer terrorists their greatest hopes for meaningful victories. But even on these abstract battlefields, from the psychological to the theological to the hypothetical, the momentum ten years after 9/11 seems to belong to their opponents.

CHAPTER FIVE

# Where Are We in the "War of Ideas"?

*Angel Rabasa*

*Angel Rabasa is a RAND senior political scientist who has written extensively about extremism, terrorism, and insurgency.*

For the United States, the threat of global terrorism defined the security environment in the first decade of the 21st century. After 9/11, the U.S. government developed a National Strategy for Combating Terrorism that declared that the 9/11 attacks were acts of war against the United States and that combating terrorism and securing the U.S. homeland were the country's top priorities. Homeland defense was not considered sufficient to safeguard the United States. Central to the Bush administration's conception of the war on terrorism was the need to carry the war to the terrorists in their own lairs. As the 9/11 Commission noted in its 2004 report, threats can develop on the other side of the earth and unleash weapons of unprecedented destructive power. Therefore, an effective counterterrorist strategy had to be global and had to enlist the full cooperation of all of the countries engaged in this war on terrorism.

But American conceptions of the global terrorist threat evolved. In the immediate aftermath of 9/11, the Bush administration defined the global war on terrorism as a struggle against international terrorism broadly defined—that is, against "premeditated, politically motivated violence perpetrated against noncombatant targets by subnational groups or clandestine agents."

By defining the threat in generic terms, the administration hoped to deflect allegations that the global war on terrorism was a war against

Islam. This vagueness was a political necessity, since the radical Islamist argument against the United States and the West was precisely that the war on terrorism was a war of civilizations, the latest iteration of the West's long war against Islam. (It is imperative to distinguish between Islam, a major world religion, and Islamism, a modern religiously based ideology that has political ends and whose most extreme followers often engage in violence.)

Although the generic definition of the war on terrorism was a political necessity, it created some conceptual confusion, because terrorism is, after all, a tactic that can be employed by different groups to pursue different objectives, irrespective of ideology. An unspecific definition of the threat that did not distinguish between groups with limited means and objectives—Basque terrorists, the Kurdistan Workers' Party, Shining Path (in Peru), or even a Muslim entity like Hamas—and terrorist groups with global reach and unlimited objectives, such as al Qaeda, could not provide an adequate framework for confronting the complex Islamist terrorist threat.

Consequently, the past decade has witnessed a series of shifting U.S. tactics on the ideological battleground, from aggressive initial advance to calculated retreat. Ironically, European allies have shifted their focus in the other direction. But the sudden eruption of democratic movements throughout the Arab world in 2011 could give the United States an opportunity to advance on the ideological front once again and to claim the strategic high ground in the fight against radical Islamism in the Muslim world.

## From the Generic to the Specific

As a practical matter, from the beginning, the focus of the global war on terrorism was not on all international terrorists, but on al Qaeda and the groups affiliated and associated with it. The 9/11 Commission made this explicit by identifying the source of the danger as Islamist terrorism—especially the al Qaeda network, its affiliates, and its ideology. As the commission's report mentions, bin Laden and other Islamist terrorist leaders draw on a long tradition of extreme intolerance within a minority stream of Islam, which is fed by grievances widely felt throughout the Muslim world. So the commission recommended that the strategy of the United States and its allies should have two goals: to dismantle the al Qaeda network and to prevail in the long term over the ideology that gives rise to Islamist terrorism.

The proposition that the United States and its allies were engaged in a war of ideas with radical Islamism gained currency within the Bush administration and more broadly among U.S. and international academic and public policy circles. As early as September 2002, the *National Security Strategy* stated that along with its military response, the United States must "wage a war of ideas" against international terrorism "to ensure that the conditions and ideologies that promote terrorism do not find fertile ground in any nation." The country would do this by "diminishing the underlying conditions that spawn terrorism" and "using effective public diplomacy to promote the free flow of information and ideas to kindle the hopes and aspirations of freedom of those in societies ruled by the sponsors of global terrorism."[1]

Effectively waging a war of ideas, however, required overcoming immense obstacles, including the prevalence of authoritarian regimes in the Middle East (some of which were key U.S. allies) and the weakness of moderate and liberal Muslim networks and institutions. A 2007 RAND Corporation study noted that although radicals are a small minority in almost all Muslim countries, they exercise disproportionate influence. Liberal and moderate Muslims, the study noted, do not have the organizational tools to effectively counter the radicals. Therefore, the creation of moderate Muslim networks is critical for providing moderates with a platform to amplify their message. Moreover, such networks could protect the moderates from extremists and sometimes also from their own governments, which repress moderates precisely because they can offer an acceptable alternative to authoritarian rule.[2]

---

[1] George W. Bush, *The National Security Strategy of the United States of America*, September 2002. As of May 27, 2011: http://georgewbush-whitehouse.archives.gov/nsc/nss/2002/

[2] Angel Rabasa, Cheryl Benard, Lowell H. Schwartz, and Peter Sickle, *Building Moderate Muslim Networks*, Santa Monica, Calif.: RAND Corporation, MG-574-SRF, 2007. As of May 24, 2011: http://www.rand.org/pubs/monographs/MG574.html

The RAND study advocated drawing on the experience of the United States and its partners during the early years of the Cold War to help build free and democratic institutions and organizations. The study offered specific recommendations on how to apply the lessons of that experience to the conditions in Muslim countries and communities today. The recommendations underscored the promise of targeting assistance to liberal and secular Muslim academics and intellectuals, young moderate religious scholars, community activists, women's groups engaged in gender-equality campaigns, and moderate journalists and writers.

In principle, the Bush administration was open to an integrated effort to contest the ideological ground with radical Islamists. The 2006 *Quadrennial Defense Review* stated that the United States was involved in a war that is "both a battle of arms and a battle of ideas" in which ultimate victory could be won only "when extremist ideologies are discredited in the eyes of their host populations and tacit supporters."[3] However, despite recognition at senior levels in the Bush administration of the ideological nature of the conflict with radical Islamism, the United States never developed a coherent policy of engagement with like-minded Muslim partners. There was no explicit U.S. policy to help build moderate Muslim networks and institutions, although some institution-building activity took place as a by-product of programs of democracy promotion, civil society development, and public diplomacy. Some potentially consequential initiatives never got off the ground. For instance, consideration was given to establishing a scholarly journal analogous to the celebrated Cold War publication *Problems of Communism* to provide a venue for a serious discussion of Islamism, but bureaucratic inertia prevented the concept from being realized before the Bush administration left office.[4]

Unlike the experience of the Cold War, when the counterideological component of the containment strategy developed by the Truman

---

[3] Department of Defense, *Quadrennial Defense Review Report*, Washington, D.C., February 6, 2006.

[4] The author was personally involved in discussions with senior Bush administration officials regarding the creation of this publication.

administration had been continued by the successor administration, the advent of the Obama administration marked a sharp break with the Bush administration's relatively modest efforts to influence the ideological debate in the Muslim world. The new administration abandoned its predecessor's "Democracy Agenda" (which, of course, had been tempered in practice by considerations of *realpolitik*, especially with regard to such key allies as Egypt) and further narrowed its definition of the adversary from "Islamist extremists" to "violent extremists"[5] or "al Qaeda and its terrorist affiliates."[6] At a speech at the Center for Strategic and International Studies in May 2010, John Brennan, the Assistant to the President for Homeland Security and Counterterrorism, said that the administration would not describe the enemy as jihadists or Islamists "because jihad is a holy struggle, a legitimate tenet of Islam meaning to purify oneself or one's community."[7] Tellingly, the U.S. Department of Defense shut down its Office of Support for Public Diplomacy, which had been responsible for coordinating Defense Department information campaigns overseas and had taken the lead on the Pentagon's part of the "war of ideas."

While the Obama administration has moved to deemphasize the ideological challenge of Islamism, Western Europe has moved in the opposite direction. For a time, some European governments were willing to recognize and even to promote Islamists, although in some cases this appears to have been less a conscious policy than an inability to distinguish Islamists from mainstream Muslims. For several years, for instance, the British government recognized the Islamist-led Muslim Council of Britain as its main interlocutor in the British Muslim community. However, since about 2005, British authorities have distanced

---

[5] Barack Obama, "Remarks by the President on a New Beginning," speech given at Cairo University, Cairo, Egypt, June 4, 2009. As of May 27, 2011: http://www.whitehouse.gov/the_press_office/Remarks-by-the-President-at-Cairo-University-6-04-09/

[6] John Brennan, "Securing the Homeland by Renewing America's Strength, Resilience, and Values," presentation at the Center for Strategic and International Studies, Washington, D.C., May 26, 2010.

[7] John Brennan, "Securing the Homeland by Renewing America's Strength, Resilience, and Values," presentation at the Center for Strategic and International Studies, Washington, D.C., May 26, 2010.

themselves from Islamist groups and have redirected their support to non-Islamists and even organizations that are strongly confronting Islamists.[8]

The discussion in Europe differs from that in the United States in that while the U.S. perception of the threat is primarily one of radicalization as a stage in the progression toward terrorism, Europeans see radicalization in the context of the broader social problem of integrating the continent's Muslim communities into the majority societies. Increasingly, the policies of engagement with Muslim communities in major European countries are directed at promoting the acceptance of the core values of their nations. As early as 2003, the Dutch security services warned about the need to address the manifestations of radicalization that, while not directly violent, fostered violence and were harmful to the democratic legal order.[9] Eight years later, in a speech at the Munich Security Conference on February 5, 2011, British Prime Minister David Cameron criticized the policy of "state multiculturalism" as a failure and blamed it for contributing to conditions of radicalization that could lead to terrorism. He went on to propose a more robust counterideological approach: "Instead of ignoring this extremist ideology," he said, "we—as governments and societies—have got to confront it."[10]

Despite the change in rhetoric from the Bush administration to the Obama administration, the underlying challenge facing the United States from Islamist extremism has not changed. The war of ideas continues under other rubrics. As the Europeans increasingly recognize, at its core the challenge of radical Islamism is ideological. Ideology shapes in important ways the structure, objectives, and strategies of radical Islamist organizations and, as such, serves as radical Islamism's "oper-

---

[8] See Angel Rabasa, Stacie L. Pettyjohn, Jeremy J. Ghez, and Christopher Boucek, *Deradicalizing Islamist Extremists*, Santa Monica, Calif., RAND Corporation, MG-1053-SRF, 2010. As of May 24, 2011: http://www.rand.org/pubs/monographs/MG1053.html

[9] Kingdom of the Netherlands, General Intelligence and Security Service, Communications Department, *Annual Report 2003*, The Hague: Ministry of the Interior and Kingdom Relations, July 2004.

[10] Oliver Wright and Jerome Taylor, "Cameron: My War on Multiculturalism," *The Independent*, February 5, 2011.

ational code." To prevail against the challenge of radical Islamism, therefore, it is necessary to address comprehensively the ideological challenge that it presents—the content or substance of the ideology, its ideological agents, networks, and means of dissemination—and to devise effective countermeasures for each of these components. Of course, a government's ability to do that will be influenced by the country's political culture and prevailing legal system. It would be difficult for the United States, for instance, to replicate some European approaches to counterradicalization.

## Whose War Is It?

The difficulty is that the United States, or other secular Western countries for that matter, can do little to contest the ideological ground with radical Islamism directly. Only Muslims themselves have the credibility to do that.[11] A key challenge, therefore, is identifying appropriate Muslim partners. Within the academic and policy communities in the United States and Europe, there is a major debate on whether Islamists should be engaged as partners or whether our partners should be the so-called moderate and liberal Muslims whose political values are congruent with the values that underlie modern liberal societies.

The argument in favor of engaging Islamists has three predicates: first, that Islamists represent the only real mass-based alternative to authoritarian regimes in the Muslim world and especially the Arab world; second, that Islamist groups such as the Egyptian Muslim Brotherhood have evolved and now support pluralistic democracy;[12] and

---

[11] Satloff advocated "tapping America's underappreciated, underutilized anti-Islamist allies" among the world's Muslims (Robert B. Satloff, *Battle of Ideas in the War on Terror: Essays on U.S. Public Diplomacy in the Middle East*, Washington, D.C.: Washington Institute for Near East Policy, 2004, p. 60). The same recommendation was made by the authors of the RAND report.

[12] As stated by Saad Eddin Ibrahim, Center for the Study of Islam and Democracy (CSID) Conference, Washington, D.C., April 22, 2005.

third, that Islamists are more likely than mainstream clerics to be successful in dissuading potential terrorists from committing violence.[13]

The argument against engaging Islamists also has three parts: First, we do not know whether the pro-democracy rhetoric and relatively more moderate discourse among Islamists represent a strategic or a tactical shift. In other words, have they ceased being Islamists, in the sense of accepting the separation of religion and the state? Or are they simply lowering the profile of one goal (the establishment of an Islamic society and state) and emphasizing a more appealing and less controversial agenda? Second, even if Islamists might be more effective in the short term in dissuading extremists from committing acts of terrorism, official recognition and support would enhance their credibility and enable them to proselytize more effectively in the community. Third, even if one concedes that in many parts of the Muslim world, moderate and liberal groups are organizationally weak and have been unable to develop substantial constituencies, if the West were to ignore these groups in favor of Islamists, it would simply perpetuate the moderates' weakness.

By and large, the arguments for partnering with Islamists are tactical—they may offer short-term benefits, if one accepts the argument that they have greater credibility among individuals at risk of violent radicalization—while the arguments for engaging non-Islamists are strategic. Providing Muslims who are supportive of modern liberal values and institutions with the tools they need to contest the ideological ground with Islamists can begin to shift the discussion within Muslim communities—as is already happening in the United Kingdom—toward a conception of Islam that is consistent with modern liberal values.

The revolutions in the Arab world in 2011 present the United States with risks and opportunities. To the extent that a political transition weakens a state's security apparatus—and sometimes the state as a whole—the state's ability to target extremists weakens. Over the short term, even in the absence of negative outcomes such as a return to authoritarianism, an Islamist takeover, or state breakdown, the upheav-

[13] This argument was made to the author by a representative of a European Foreign Ministry.

als in the Middle East may deprive the United States of partners able or willing to cooperate in the struggle against violent extremism. A future Egyptian government, for instance, would likely be constrained by what will probably be a larger role for the Muslim Brotherhood in Egypt's domestic politics.

Over the long term, however, if the democratic transitions endure, the emergence of more-inclusive and more-responsive governments in the region might provide the United States with the opportunity to reengage in the war of ideas. If, as was argued in a 2004 RAND study, the widespread failure of post-independence political and economic models in the Middle East and the ills and pathologies produced by this failure have generated much of the extremism that has concerned us since 9/11,[14] the movement toward democracy and accountable governments might take the wind out of the sails of Islamist extremist ideology and, finally, begin to drain the swamp of terrorism.

The outcome will depend on the ability of the countries undergoing a democratic revolution to sustain the process. Some may argue that the prospects that democracy will take root in countries with no history of democracy, weak civil societies, poor civil-military relations, and ethnic or religious tensions are poor. But the outbreak of democratic revolutions across the Middle East suggests that countries in the region have reached a level of development where autocracies are no longer acceptable to broad sectors of the population. The modern history of East and Southeast Asia suggests that liberal democracies can develop and flourish in countries with little previous democratic experience (even when democratization has occurred after—or was punctuated by—various returns to authoritarianism). The countries of the Greater Middle East may prove to be no exception.

---

[14] Angel Rabasa, Cheryl Benard, Peter Chalk, C. Christine Fair, Theodore W. Karasic, Rollie Lal, Ian O. Lesser, and David E. Thaler, *The Muslim World After 9/11*, Santa Monica, Calif.: RAND Corporation, MG-246-AF, 2004. As of May 24, 2011: http://www.rand.org/pubs/monographs/MG246.html

## Related Reading

Aboul-Enein, Youssef, *Militant Islamist Ideology: Understanding the Global Threat*, Annapolis, Md.: Naval Institute Press, 2010.

Albright, Madeleine K., and Vin Weber, *In Support of Arab Democracy: Why and How*, New York and Washington, D.C.: Council on Foreign Relations, Independent Task Force Report No. 54, 2005.

Charfi, Mohamed, *Islam and Liberty: The Historical Misunderstanding*, London and New York: Zed Books, 2005.

Habeck, Mary, *Knowing the Enemy: Jihadist Ideology and the War on Terror*, New Haven and London: Yale University Press, 2006.

Maher, Shiraz, and Martyn Frampton, *Choosing Our Friends Wisely: Criteria for Engagement with Muslim Groups*, London: Policy Exchange, 2009.

McMillan, Joseph, ed., *"In the Same Light as Slavery": Building a Global Antiterrorism Consensus,* Washington, D.C.: Institute for National Strategic Studies, National Defense University, 2006.

Rabasa, Angel, Peter Chalk, Kim Cragin, Sara A. Daly, Heather S. Gregg, Theodore W. Karasik, Kevin A. O'Brien, and William Rosenau, *Beyond al-Qaeda: Part 1, The Global Jihadist Movement*, Santa Monica, Calif.: RAND Corporation, MG-429-AF, 2006. As of May 24, 2011: http://www.rand.org/pubs/monographs/MG429.html

———, *Beyond al-Qaeda: Part 2, The Outer Rings of the Terrorist Universe,* Santa Monica, Calif.: RAND Corporation, MG-430-AF, 2006. As of May 24, 2011:
http://www.rand.org/pubs/monographs/MG430.html

CHAPTER SIX

# Al Qaeda's Propaganda: A Shifting Battlefield

*Eric V. Larson*

*A RAND senior policy researcher, Eric Larson recently has led studies of al Qaeda's jihadi strategy, ideology, propaganda, and discourse, and of public support for terrorism and insurgency.*

At the heart of al Qaeda's effort to build a violent social movement based on its transnational ideology of salafi jihadism—a violent fundamentalist form of Islamism—is a contest over the true nature of Islam: whether Islam is merciful, compassionate, and tolerant, imposing substantial constraints on the permissibility of violent jihad, which is the view of most mainstream Muslim thought, or whether Islam is intolerant and permissive of violent jihad, in accordance with al Qaeda's reading.[1] This contest has aptly been described both as a civil war

---

[1] As the term *jihad* literally translates to "struggle" and can take on many forms (e.g., a Muslim's personal struggle to be closer to Allah, jihad of the tongue, jihad of the pen), we use the term "violent jihad" to differentiate this concept from other forms of struggle.

Salafis are a relatively small Sunni Islamist current, largely centered in Saudi Arabia and the core Arab region. Salafis aim to practice the ascetic and puritanical form of Islam that existed under the "Rightly-Guided Caliphs"—Abu Bakr, 'Umar, 'Uthman, and 'Ali—who led the early Muslim community following the death of the Prophet Muhammad. Salafi practice is typified by the view that the Holy Koran and the traditions and sayings of the Prophet Muhammad (the *sunnah* and *hadith*) are the only true and authentic sources for Islamic belief and practice, and Salafis take a highly literalistic reading of these texts, generally evidencing ambivalence toward independent analysis and reinterpretation of their meaning (*tafsir*). Salafism includes pietists who generally eschew worldly action, those who engage in political activity but eschew violence (e.g., the Sahwists, or "Awakening"), and *salafi-jihadists* who embrace violent armed struggle. For a detailed characterization of the Salafi

within Islam itself[2] and as being analogous to the West's own century-long Reformation and Counter-Reformation of the 16th and 17th centuries.[3]

Al Qaeda's two most senior leaders asserted that propaganda and media activities are essential weapons in their efforts to influence the outcome of this "framing contest" over the heart of Islam. In an undated letter to Taliban leader Mullah Mohammad Omar, Osama bin Laden quantified the importance of propaganda and media activities: "It is obvious that the media war in this century is one of the strongest methods; in fact, its ratio may reach 90 percent of the total preparation

---

movement, see Quintan Wiktorowicz, "Anatomy of the Salafi Movement," *Studies in Conflict & Terrorism*, Vol. 29, 2006, pp. 207–239.

In one of its reports, the International Crisis Group provided compact definitions of a number of related terms, as follows:

> In the usage adopted by ICG, "Islamism" is Islam in political rather than religious mode. "Islamist movements" are those with Islamic ideological references pursuing primarily political objectives, and "Islamist" and "Islamic political" are essentially synonymous. "Islamic" is a more general expression, usually referring to Islam in religious rather than political mode but capable, depending on the context, of embracing both.

See International Crisis Group, *Saudi Arabia Backgrounder: Who Are the Islamists?* September 21, 2004, p. i. As of May 27, 2011: http://www.crisisgroup.org/en/regions/middle-east-north-africa/iran-gulf/saudi-arabia/031-saudi-arabia-backgrounder-who-are-the-islamists.aspx

For a detailed analysis of al Qaeda Central through the lens of social movement theory, see Paul K. Davis, Eric V. Larson, Zachary Haldeman, Mustafa Oguz, and Yashodhara Rana, *Understanding and Influencing Public Support for Insurgency and Terrorism*, Santa Monica, Calif.: RAND Corporation, forthcoming.

[2] As Doran writes, after 9/11 "Washington had no choice but to take up the gauntlet, but it is not altogether clear that Americans understand fully this war's true dimensions. The response to bin Laden cannot be left to soldiers and police alone. He has embroiled the United States in an intra-Muslim ideological battle, a struggle for hearts and minds in which al Qaeda had already scored a number of victories." See Michael Scott Doran, "Somebody Else's Civil War," *Foreign Affairs*, January/February 2002, pp. 22–23.

[3] See, for example, Ali Eteraz, "An Islamic Counter-Reformation," *The Guardian*, October 2, 2007. See also Khaled Abou El Fadl, *The Great Theft: Wrestling Islam from the Extremists*, San Francisco: HarperSanFrancisco, 2005, especially Part One, "The Battleground for Faith."

for the battles."[4] Al Qaeda's second in command, Ayman al-Zawahiri, ascribed a lower but still decisive level of importance to these activities in a letter to Abu Musab al-Zarqawi, the brutal al Qaeda henchman of Iraq, in July 2005: "Despite all of this, I say to you: that we are in a battle, and that more than half of this battle is taking place in the battlefield of the media."

Although they differ somewhat in the relative weight they accord to the importance of propaganda and media activities—for bin Laden, 90 percent; for al-Zawahiri, at least 50 percent—both view these activities as crucial to waging the war over Islam and to establishing the conditions under which violent jihad is permissible or impermissible. Their views, moreover, are widely shared across the jihadi community.[5]

Ironically, however, al Qaeda has not fared well in its own framing contest, thus opening the door for the movement's opponents to shape the outcome. The harms that have befallen the Muslim world as a result of al Qaeda's practice of violence have led to severe criticisms from outside the salafi-jihadi movement and from within the movement as well, and most of the available public opinion survey data suggest a decline in support for al Qaeda from within the Muslim world.

The principal implication for U.S. policy is as follows: The best strategy is to pursue actions that erode the persuasiveness of al Qaeda's narrative while avoiding actions that play into the narrative and that impede al Qaeda's ongoing self-destruction. In other words, some of the best U.S. actions would be to broker a broader settlement between Israel and the Palestinians, to promote democratic values, and to provide humanitarian assistance. Conversely, some of the worst U.S.

---

[4] Combating Terrorism Center, translation of "Letter to Mullah Mohammed 'Omar from Osama Bin Laden," CTC Harmony Document Database, AFGP-2002-600321, June 5, 2002. As of June 22, 2011: http://www.ctc.usma.edu/wp-content/uploads/2010/08/AFGP-2002-600321-Trans.pdf

[5] On these "framing contests," see Quintan Wiktorowicz, "Framing Jihad: Intramovement Framing Contests and al-Qaeda's Struggle for Sacred Authority," *International Review of Social History*, Vol. 49, 2004, pp. 159–177; and Quintan Wiktorowicz, "A New Approach to the Study of Islamic Activism," *IIAS Newsletter*, Vol. 33, March 2004. As of June 22, 2011: http://www.iias.nl/iiasn/33/RR_IA.pdf

actions would be to deploy additional combat forces in Muslim lands and to interject an American voice into the internal theological debate over the nature of Islam. In short, success on the ideological battlefield will depend more on efforts to promote American ideals than on American might.

## Al Qaeda Under Attack

In recent years, al Qaeda has come under attack by leading Muslim voices in the form of criticism of its theological, jurisprudential, and strategic reasoning. The criticism has been particularly strident regarding al Qaeda's violence against fellow Muslims. Among the harshest attacks coming from outside the salafi-jihadi movement are those from Saudi cleric Salman al-Awda and Egyptian cleric Yusuf al-Qaradawi.

In a September 14, 2007, open letter titled "A Ramadan Letter to Osama bin Laden," al-Awda asked bin Laden: "How much blood has been spilled? How many innocent children, women, and old people have been killed, maimed, and expelled from their homes in the name of 'al Qaeda'?"[6] Al-Awda reportedly had been an important early influence on bin Laden's religious views.[7] In June 2009, al-Qaradawi, a highly influential Qatar-based cleric who is chairman of the International Federation of Muslim Scholars, published a book in which he repudiated al Qaeda's concept of jihad as a "mad declaration of war upon the world."[8] This work, by a popular, mainstream Islamic cleric

[6] Shaykh Salman Bin Fahd al-Awda, "A Ramadan Letter to Osama bin Laden," published on his Islam Today website.

[7] Al-Awda was imprisoned for five years in the 1990s for his Sahwist opposition to the Saudi government and was released in 1999 upon being judged "rehabilitated." He has since become a very prominent and popular mainstream cleric with a television program and website.

[8] Yusuf al-Qaradawi, *Fiqh al-Jihad: A Contemporary Study of Its Rules and Philosophy in the Light of the Qur'an and Sunna*, 2009, serialized in seven parts beginning with Mahir Hasan, "Al-Misri Al-Yawm Publishes Revisions of Jihad by Al-Qaradawi: '(1) Three Groups in the Nation with Different Concepts of Jihad: The First Rejects It, the Second Declares War on the World, and the Third Adopts a Middle-of-the-Road Approach,'" Al-Misri Al-Yawm, June 29, 2009.

who has a weekly program on al-Jazeera television and who retains a storied ability to serve as a barometer of broader mainstream Muslim opinion, suggested that the mainstream tide might have turned against al Qaeda.[9]

In many ways, the attacks on al Qaeda and "revisions" of jihadi doctrine from within the salafi-jihadi movement have been even more scathing, and there should be little doubt that these high-profile defections from the movement have raised doubts about al Qaeda among its cadres and sympathizers. In April 2007, the Kuwaiti salafi-jihadi scholar Hamid al-Ali issued a *fatwa* (religious ruling) against the establishment of the Islamic State of Iraq, implicitly criticizing al Qaeda's affiliate in Iraq for its violence. "The spreading of bigotry and rancor, even if wrapped in the cloak of religion, is the work of the devil and of people who follow their own caprice. It must be avoided. Everyone must keep distance from such a dangerous path," wrote al-Ali.[10]

Late 2007 saw the release of a book by Sayyid Imam Abd al-Aziz al-Sharif, the former Egyptian Islamic Jihad Organization ideologue and author of a classic jihadi manual. In his book, Sayyid Imam extensively revised his earlier positions on the jurisprudence of jihad, making violent jihad impermissible under most circumstances.[11] This was an especially important attack from within the salafi-jihadi camp,

---

[9] George Washington University professor Marc Lynch described Qaradawi as "probably the single most influential living Sunni Islamist figure." See Marc Lynch, "Qaradawi's Revisions," July 9, 2009. As of June 19, 2011: http://lynch.foreignpolicy.com/posts/2009/07/09/qaradawis_revisions

[10] Hamid al-Ali fatwa, April 2007. Al-Ali has a more nationalist salafi-jihadi orientation, compared with al Qaeda's transnational perspective.

[11] The book was serialized in the Cairo newspaper *Al-Misri Al-Yawm* beginning on November 18, 2007, with the article "Exclusive, Al-Misri Al-Yawm Publishes the Theological Revisions of Jihad. 'Jihad for Allah's Sake' Involved Religious Violations, Most Importantly Killings on the Basis of Nationality, Skin Color, and Denomination: First Part of Sayyid Imam's "Rationalizing Jihad in Egypt, the World," *Al-Misri Al-Yawm*, November 18, 2007. Sayyid Imam followed up the release of his book with a six-part interview in the London paper *Al-Hayah* in December 2007 that began with "Al-Hayah in Egypt's Turrah Prison Interviews Author of the Document 'The Rationalization of Jihad in Egypt and the World.' Dr Fadl: 'Al-Zawahiri Deceived Me and Was the Reason I Was Accused in Al-Sadat Case. I Left Jama'at al-Jihad After IT INSISTED on Operations Inside Egypt and Distorted My Book, A Compilation'," *Al-Hayah*, December 8, 2007.

as Sayyid Imam had formerly been the ideological mentor to Ayman al-Zawahiri.[12] Sayyid Imam's *volte-face* on the permissibility of jihad sent shock waves through the salafi-jihadi community and led to a public dispute with al-Zawahiri. Also in late 2007, former Libyan Islamic Fighting Group leader Nu'man Bin 'Uthman issued his opening salvo against al Qaeda, framing his criticisms in both strategic and jurisprudential terms.[13,14]

In January 2009, the Egyptian Islamic Group—a prominent former jihadi group, most of whose leadership declared a unilateral cease-fire in the late 1990s and formally renounced violence in March 2002—issued a statement urging al Qaeda to observe a cease-fire until it could assess the intentions of the recently elected Obama administration.[15] In September 2009, the Libyan Islamic Fighting Group released a new "code" for jihad in the form of a 417-page religious document

---

[12] Sayyid Imam (also known as Abd Al-Qader Bin 'Abd Al-'Aziz, or Dr. Fadl) was the Egyptian Islamic Jihad's leading ideologue and the author of a classic jihadi text called "The Essentials of Making Ready [for Jihad]" that initially was used by al Qaeda in its training program. Sayyid Imam's revisions were subsequently denounced in Ayman al-Zawahiri's book, *The Exoneration*. For a summary, see Lawrence Wright, "The Rebellion Within: An Al Qaeda Mastermind Questions Terrorism," *The New Yorker*, June 2, 2008. As of February 2011: http://www.newyorker.com/reporting/2008/06/02/080602fa_fact_wright

[13] See "Former Libyan Fighting Group Leader Responds to the Announcement That His Group Has Joined Al-Qa'ida. Bin-Uthman to Al-Zawahiri: Dissolve 'the Islamic State of Iraq' and Halt Your Operations in Both Arab and Western Countries, *al-Hayah* (London), November 7, 2007; Nu'man Bin 'Uthman, "Al-Qaeda: Your Armed Struggle Is Over," open letter to Usama Bin Laden, September 2010. As of January 2011: http://www.quilliamfoundation.org/index.php/component/content/article/690; and *A Translation of* The Other Face of Al-Qaeda *by Camille Tawil*, translation by Maryam El-Hajbi and Mustafa Abdulhimal, Quilliam Foundation, November 2010. As of March 2011: http://www.quilliamfoundation.org/index.php/component/content/article/728

[14] There is some disagreement about how significant some of these renunciations are in terms of their impact on potential recruits or those already radicalized.

[15] See Abd-al-Sattar Hatitah, "*Al-Jama'ah al-Islamiyah* [the Islamic Group] calls on Al-Qa'ida to Declare a 4-Month Truce with the West 'to Test Obama's Intentions'; the group's theorist told Al-Sharq al-Awsat, "We fear that the organization might carry out attacks," *Al-Sharq al-Awsat*, January 24, 2009.

that served as an extended critique and recantation of al Qaeda's reading of the jurisprudence of jihad.[16]

In November 2010, former al Qaeda spokesman Suleman Abu Ghaith released a book that constituted another damning indictment of al Qaeda from a former insider. Echoing other critics, Abu Ghaith emphasized that after having pledged allegiance to Taliban leader Mullah Mohammed Omar while in Afghanistan, bin Laden violated Islamic law by failing to abide by Omar's instructions not to attack the United States. This violation, according to Abu Ghaith and others, made bin Laden an unfit leader who deserved punishment.[17]

Finally, in late 2010, al Qaeda's former military planner Sayf al-Adl wrote a number of letters calling upon al Qaeda's leaders to conduct a comprehensive review of their operations, including the 9/11 attacks, for the purpose of "assessing the past stage, learning the lessons, and drawing up a strategy for the future."[18] The letters were highly critical of al Qaeda's past mistakes, but it is also alarming that al-Adl

---

[16] See Quilliam Foundation, *A Selected Translation of the LIFG Recantation Document*, translation by Mohammed ali Musawi. As of March 2011: http://www.quilliamfoundation.org/acatalog/Books.html

[17] Both Abu Ghaith and the late al Qaeda third-in-command Abu Hafs are believed to have opposed the 9/11 attacks in part because they constituted a breach of shariah obligations, as just described.

[18] See Jamal Isma'il, "Signs of Split in 'Al-Qa'ida' Organization: 'Sayf-al-Adl' Launches Revisions Against Violence," *Al-Hayah* (London), January 11, 2011; and Rajeh Said, "Former al-Qaeda Spokesman Criticises 'Culture of Killing' and 'Monopoly of Jihad,'" *Al-Shorfa*, December 16, 2010. As of February 2010: http://www.al-shorfa.com/cocoon/meii/xhtml/en_GB/features/meii/features/main/2010/12/16/feature-01

In an early November 2010 letter to Cairo newspaper *Al-Yawm Al-Sabi*, al-Adl denied any remaining ties to al Qaeda, but conflicting, unconfirmed reports suggest that he may now be in North Waziristan and in command of al Qaeda's international operations. See Mahmud Al-Mamluk, "We Publish Exclusively a Letter Received from the Principal Suspect in the 'Bomb Parcels' Case. Brigadier Makkawi: 'I Have No Relationship with Al-Qa'ida and I Cannot Leave Pakistan Because of the Ban Imposed by the American Intelligence,'" *Al-Yawm Al-Sabi*, November 9, 2010, p. 6; Syed Saleem Shahzad, "Parcel Bombs Point to New al-Qaeda Plans," *Asia Times*, November 3, 2010; Syed Saleem Shahzad, "Broadside Fired at al-Qaeda Leaders," *Asia Times*, December 10, 2010. Al-Adl is believed by many analysts to have been the author of a 20-year al Qaeda strategy for establishment of an Islamic Caliphate that is described in Fouad Hussein, *Al-Zarqawi: The Second Generation of Al Qaeda*, in Arabic, 2005. For a highly readable discussion, see Lawrence Wright, "The Master Plan: For the New Theorists of Jihad,

appeared to be advocating that the group emphasize a strategy focused on attacking the United States rather than local Muslim regimes.[19] In the wake of bin Laden's death, al-Adl was appointed al Qaeda Central's acting chief until al Qaeda's Shura Council could choose a more permanent successor.[20]

## Defending the Movement

In response to these developments, al Qaeda Central has felt compelled to mount a propaganda counteroffensive against its critics, with al-Zawahiri taking the lead.

Between December 2007 and January 2008, al Qaeda's accredited web forums accepted questions for an "open interview" with al-Zawahiri in which, it was promised, he would address any and all questions directed to him.[21] His response took the form of two audio recordings released in April 2008, which were noteworthy for the number of questions regarding al Qaeda's theological and jurisprudential justifications for causing so many Muslim deaths—and for al-Zawahiri's evasive and inconsistent responses to these questions.[22] It is rather difficult to

---

Al Qaeda Is Just the Beginning, *The New Yorker*, September 11, 2006. As of March 2011: http://www.newyorker.com/archive/2006/09/11/060911fa_fact3

According to Fouad Hussein, a number of other figures, including Assad al-Jihad, may have had a more prominent role in developing this strategy.

[19] For an excellent analysis of al-Adl's letters, see Vahid Brown, "Al-Qa'ida Revisions: The Five Letters of Sayf al-'Adl." February 10, 2011. As of April 2011: http://www.jihadica.com/al-qa%E2%80%99ida-revisions-the-five-letters-of-sayf-al-%E2%80%98adl/

[20] Javed Rashid, "Saifal Adel Made Acting Al-Qaeda Chief," *The News Online*, Islamabad, May 17, 2011.

[21] Judged by the pointed questioning, it appears that the questions were reported verbatim as they were submitted.

[22] See Al-Sahab Media Production Institute, "Al-Zawahiri Answers Questions by Al-Qa'ida Critics, Criticizes Al-Qaradawi, HAMAS," April 3, 2008. As of January 2011: http://triceratops.brynmawr.edu/dspace/handle/10066/4675

See also Al-Sahab Media Production Institute, "Jihadist Website Releases 'Second Round' of Open Interview with Al-Zawahiri," April 23, 2008. As of January 2011: http://triceratops.brynmawr.edu/dspace/handle/10066/4677

view al-Zawahiri's "open interview" as anything but a public relations disaster. In March 2008, al-Zawahiri responded to Sayyid Imam's book-length attack with a book of his own.[23] In February 2009, Sayyid Imam released a second book, attacking al-Zawahiri's book.[24] Then in early 2011, perhaps feeling backed into a corner, both al-Zawahiri and Shaykh Atiyatallah, another senior al Qaeda ideologue, released statements emphasizing the importance of *avoiding* deaths to innocent Muslims and non-Muslim bystanders.[25]

Nonetheless, al Qaeda Central's house theologian, Abu Yahya al-Libi, also has been enlisted to justify killing innocent Muslims and "spies." In 2006, he expounded upon a relatively obscure jurisprudential concept called *al-tatarrus* (human shields) that allowed for deaths of innocent Muslims if they were being used by enemies as human shields.[26] In June 2009, al Qaeda's al-Fajr Media Center released a book by al-Libi that sought to justify punishing fellow Muslims who provided intelligence on al Qaeda. This development was widely viewed as evidence of al Qaeda Central's growing preoccupation with personal

---

[23] A translation of al-Zawahiri's book by the NEFA Foundation was available at http://www.globalterroralert.com/al-qaida-leaders/40-dr-ayman-al-zawahiri-qthe-exonerationq.html, as of February 2011. Among other lines of attack, al-Zawahiri sought to delegitimize Sayyid Imam's arguments by pointing to the fact that the author was currently being held in an Egyptian prison and that the revisions should be viewed as an effort to escape further torture, or secure release.

[24] See Kamal Habib, "Another Wave of Jihadist Adjustment: Internal Debates of the Movement," ArabInsight.org, March 2009. As of February 2011: http://www.arabinsight.org/aiarticles/216.pdf

Habib is a founding member and former leader of the Egyptian Islamic Group.

[25] The first of Al-Zawahiri's series of messages, titled "Message of Hope and Glad Tidings to Our People in Egypt," was released on February 18, 2011, and Shaykh Atiyatallah's message, titled "Glorifying the Sanctity of Muslim Blood" was released on February 23, 2011. In earlier private communications to then–al Qaeda in Iraq leader Abu Mus'ab al-Zarqawi, both al-Zawahiri and Atiyatallah had also criticized al-Zarqawi for killing fellow Muslims, arguing that such actions were eroding Muslims' support for al Qaeda.

[26] See Jack Barclay, "Al-Tatarrus: al-Qaeda's Justification for Killing Muslim Civilians," *Terrorism Monitor*, Vol. 8, Issue 34, September 9, 2010. As of February 2011: http://www.jamestown.org/programs/gta/single/?tx_ttnews[tt_news]=36803&cHash=a672f81bc8

Al Qaeda appears to have employed the concept of *al-tatarrus* to justify civilian deaths at least since the Riyadh bombings in May 2003.

security, but it also should have been viewed as a preemptive jurisprudential defense of a possible campaign of assassination.[27]

Perhaps in recognition of the lack of stature that al-Zawahiri, al-Libi, and other ideologues have on religious matters outside of their movement, al Qaeda Central leaders also have increasingly called for support from other "honest" or "truthful" scholars to assist in the ideological battle against al Qaeda's critics.[28] While some scholars, such as Anwar al-Awlaki and Mansur al-Shami, have answered the call, the strong impression is that their stature is nowhere near that of those in the growing roster of al Qaeda's critics and enemies.

The conventional wisdom typically has it that al Qaeda's jihadi propaganda and media activities are hugely successful within the Muslim world and that al Qaeda is dominating the "information war," humbling America's own meager capabilities to influence Muslim attitudes.[29] To be sure, al Qaeda's propaganda and media strategy benefit from its ability to employ various symbols and slogans of Islam and Islamism in support of its program; for instance, al Qaeda's leaders have identified themes—the liberation of Palestine being the preeminent one—that find resonance at a deeply emotional level for much of

---

[27] See, for example, "Guidebook on Muslim Spies Reveals Qaeda's Fear," *al-Arabiyah*, July 9, 2009.

[28] See, for example, al-Zawahiri's "Six Years Since Invasion of Iraq and 30 Years Since Signing Peace Agreement with Israel," al-Sahab Media Production Establishment, April 20, 2009, in which al-Zawahiri stated:

> I particularly address my appeal to the honest scholars of the nation, who declared and are declaring the truth; who exposed the facts to the nation and distinguished between the traitors who are selling the nation to the Crusaders and the Jews and the true mujahidin, who sacrifice their souls, money, and everything they have for the sake of the nation's glory and dignity.
>
> I ask the true scholars of the nation to expose the traitor merchants of religion, who cooperated with the Crusader occupation in Iraq and Afghanistan; who have not thus far released even a single fatwa about the need to fight the Crusaders in Iraq and Afghanistan. So how can they release a fatwa against those who brought them to power with their weapons?

[29] For a relatively recent—and somewhat overblown—report on al Qaeda's information offensive, see Craig Whitlock, "Al-Qaeda's Growing Online Offensive," *The Washington Post*, June 28, 2008.

the Muslim world.[30] However, al Qaeda's ideological and propaganda weaknesses are more apparent than their strengths.

Al Qaeda's leaders and propagandists view themselves as being on the defensive in a media war in which they are vastly overmatched by mass media organizations that are in thrall to the Crusader United States, Zionist Israel, and local apostate regimes.[31] Al Qaeda's Internet-based propaganda and media production and distribution capabilities thus reflect an effort to overcome the organization's great weaknesses both in competing with the well-financed public diplomacy, propaganda, and other informational capabilities of Western and Muslim governments and in gaining access to those same Western and Muslim mass media, which it views as inherently hostile to its cause.[32] Even pan-Arab media stalwarts such as al-Jazeera and al-Arabiyah are viewed as enemies of the movement.[33]

Thus, in its "framing contests" with those who promote a more tolerant view of Islam, al Qaeda appears to be vastly outgunned. More-

---

[30] For example, in his 2001 book *Knights Under the Prophet's Banner*, al Qaeda second-in-command Ayman al-Zawahiri identifies the issue of Palestine as the topic that is most likely to find resonance in the larger Muslim world, and bin Laden devoted a March 14, 2009, statement to outlining "Practical Steps for the Liberation of Palestine," in large measure to counter the growing perception at the time that al Qaeda was less relevant to developments in Palestine than Hamas. That al Qaeda has cynically, and somewhat unconvincingly, manipulated the Palestinian issue is without dispute.

[31] In August 2008, for example, the Global Islamic Media Front released a book by Muhammad Bin-Zayd al-Muhajir detailing "The Media War Against People of Islam."

[32] In his July 9, 2005, letter to al Qaeda in the Land of Two Rivers, Emir Abu Mus'ab al-Zarqawi, al-Zawahiri decried the "malicious, perfidious, and fallacious campaign by the deceptive and fabricated media." See Ayman al-Zawahiri, Letter for Abu Musab al-Zarqawi, in Arabic, July 9, 2005, originally posted at http://www.dni.gov

[33] For example, al-Zawahiri notes in the second edition of his book *Knights Under the Prophet's Banner* that "the media and propaganda bodies in our countries are in the hands of groups that do not favor the Islamists. Accordingly, the right of publishing is controlled by them, to the exclusion of the Islamists. In doing so, they are following the footsteps of the West, where Jews are in control of the media and the propaganda bodies." In October 2007, al-Jazeera's broadcast of a heavily edited bin Laden speech led to widespread condemnation of the network by al Qaeda–affiliated jihadi web forum members, and in August 2008, the al Qaeda–accredited web forum *al-Hisbah* expelled the pan-Arab al-Arabiyah Channel from further participation in the forum as a result of its alleged "lies and distortions."

over, the fact that al Qaeda Central figures are being forced to spend so much time deliberating over how best to demonstrate their theological and jurisprudential bona fides and defending themselves against attacks by detractors within their own theological circles is a highly encouraging sign.

## Implications for U.S. Strategy and Policy

Suspicions of U.S. intentions run deep in many parts of the Muslim world, and it can be difficult to overcome the undercurrent of sheer resentment of U.S. superpower status, whatever Washington does, but al Qaeda's current situation nonetheless presents opportunities, especially now, given the death of the charismatic and organizationally adept bin Laden and the opening of political space in some Muslim countries as a result of the Arab Spring. With al Qaeda already under considerable pressure from within the Muslim community, what, if any, fruitful role can the United States play in what is essentially an intra-Muslim debate about the nature of Islam and the permissibility of violent jihad? In short, U.S. strategy and policy should take the long view, which means focusing on three objectives: (1) reducing or eliminating the irritants that fuel support for al Qaeda, (2) promoting universal democratic and humanitarian values, and (3) avoiding actions and rhetorical missteps that reinforce al Qaeda's narrative.

**Reducing or eliminating the irritants that fuel support for al Qaeda.** Al Qaeda's grand strategy is based on the dual assumptions that attacks on the United States can lure it into intervening militarily in Muslim lands and that Muslims can then more easily be mobilized into jihad beneath al Qaeda's banner. With a withdrawal of U.S. military forces from Iraq by the end of 2011 and substantial reductions of forces in Afghanistan envisioned to begin by the end of 2014, this irritant will diminish, and al Qaeda will find it increasingly difficult to exploit this issue in mobilizing jihadists.[34] It bears mention-

[34] The November 2010 NATO Lisbon Summit endorsed Afghan forces assuming "full responsibility for security across the whole of Afghanistan" by the end of 2014. The Afghans,

ing that bin Laden's own radicalization appears to have been fueled by the 1991 Gulf War, the deprivations experienced by Iraqis in the war's aftermath, and the prolonged presence of U.S. troops in Saudi Arabia following that war. Future decisions to deploy or permanently station U.S. military forces in Muslim lands must be carefully weighed against their potential consequences for fueling recruitment and mobilization into al Qaeda extremism. In a similar vein, the United States can weaken al Qaeda's ability to exploit the Palestinian issue by continuing efforts to expand the writ and influence of the Palestinian Authority and to promote a broader settlement between Israel and the Palestinians. Al Qaeda has judged that the Israeli occupation of Palestine is the issue that is most salient to Muslims worldwide—and the most easily exploited in its propaganda and recruitment efforts. Thus, despite the current impasse in Israeli-Palestinian negotiations, it is crucial that the United States continue its decades-long effort to promote a broader settlement.

**Promoting universal democratic and humanitarian values.** By promoting and supporting democratic reform in the Muslim world,[35] the United States aligns itself with the aspirations of most ordinary Muslims and is better positioned to marginalize al Qaeda because of the organization's outright rejection of democracy and peaceful political competition.[36] The overthrow of corrupt and despotic leaders and

---

however, have suggested that Afghan forces will assume the lead on security in all areas of Afghanistan by the end of 2014, which suggests that some NATO ISAF forces might remain beyond 2014.

[35] For example, the Egyptian revolution reportedly benefited from earlier U.S. investments in training the young Egyptians who led the uprising, and in June 2011, the G-8 pledged $20 billion to support political, social, and economic reform in so-called "Arab Spring" nations.

[36] Al Qaeda's theorists view democracy and politics as a form of polytheism in the sense that democracy and secular politics place man-made institutions between man and God's rule and are therefore destructive of true monotheism (*tawhid*). Writing of the influential Jordanian salafi-jihadi ideologue Abu Muhammad al-Maqdisi's views toward democracy, for example, Brooke writes:

> Since democrats and citizens of Arab constitutional monarchies pledge loyalty to these humanly-devised laws, Maqdisi finds them guilt of idolatry, as they worship a false deity or a *taaghut* [worshipping something other than Allah] rather than worshipping

the opening of political space, as occurred in the recent revolutions in Tunisia, Egypt, and Libya, can help sap support for al Qaeda's extremism and channel energies toward peaceful political competition. Efforts by the United States to secure the release of political prisoners also can provide tangible signs of the U.S. commitment to democratic values. Finally, by providing humanitarian assistance and disaster relief, as it did following the December 2004 Indian Ocean tsunami that struck Indonesia and other nations in the region and the October 2005 earthquake that struck Pakistan, the United States can soften its image abroad, build goodwill in Muslim nations, and help to inoculate their populations against al Qaeda propaganda and rhetoric promoting violence against America.

**Avoiding actions and rhetorical missteps that reinforce al Qaeda's narrative.** U.S. efforts to directly influence intra-Muslim debates over the nature of Islam and the permissibility of violent jihad seem highly unlikely to be effective. The theological and jurisprudential conditions and constraints that bear on the conduct of violent jihad are complex and subtle, and official U.S. efforts to opine on or influence such matters carry grave risks of both alienating potential friends and allies within the Muslim world and reinforcing al Qaeda's narrative that the United States aims to refashion Islam into a moderate or even secular form. U.S. deeds of the types outlined above matter far more than words, but U.S. words can also cause great harm. Characterizing U.S. efforts against al Qaeda as part of a "crusade" or clash of civilizations may play well with some domestic audiences, but ultimately, it is likely to reinforce al Qaeda's narrative of an Islam that is under attack by the non-Muslim world.

This is clearly not a strategy for quick, short-term results. Rather, as was the case with the earlier U.S. sustained effort to promote democracy, markets, and individual freedom over totalitarian communism,

---

> Allah alone. As such, they have been corrupted by polytheism, and have forsaken the *tawhid* [monotheism] of Allah.

See Steven Brooke, "The Preacher and the Jihadi," in Hillel Fradkin, Husain Haqqani, and Eric Brown, eds., *Current Trends in Islamist Ideology*, Vol. 3, Washington, D.C.: Hudson Institute, February 2006, pp. 52–66.

it is a strategy that is likely to require decades or generations of effort. However, only by taking the long view will the United States be able to establish conditions that favor the triumph of a merciful, compassionate, and tolerant Islam over the violent nihilism of al Qaeda and its fellow travelers.

## Related Reading

Brown, Vahid, *Cracks in the Foundation: Leadership Schisms in al-Qa'ida from 1989–2006*, West Point, N.Y.: Combating Terrorism Center, January 2007. As of June 2, 2011:
http://www.ctc.usma.edu/posts/cracks-in-the-foundation-leadership-schisms-in-al-qaida-from-1989-2006

Hegghammer, Thomas, "Jihadi Studies: The Obstacles to Understanding Radical Islam and the Opportunities to Know It Better," *The Times Literary Supplement* (London), April 4, 2008, pp. 15–17.

Moghadam, Assaf, and Brian Fishman, *Self-Inflicted Wounds: Debates and Divisions Within al-Qa'ida's and Its Periphery*, West Point, N.Y.: Combating Terrorism Center, December 2010. As of June 2, 2011:
http://www.ctc.usma.edu/posts/self-inflicted-wounds

Rollins, John, *Al Qaeda and Affiliates: Historical Perspective, Global Presence, and Implications for U.S. Policy*, Washington, D.C.: Congressional Research Service, R41070, January 25, 2011.

———, *Osama bin Laden's Death: Implications and Considerations*, Washington, D.C.: Congressional Research Service, R41809, May 5, 2011.

Shahzad, Syed Saleem, *Inside al-Qaeda and the Taliban: Beyond Bin Laden and 9/11*, London: Pluto Press, 2011.

CHAPTER SEVEN

# Have We Succumbed to Nuclear Terror?

*Brian Michael Jenkins*

*Brian Michael Jenkins is a senior adviser to the president of the RAND Corporation.*

We live at the edge of doom. President Obama has declared that "the single biggest threat to U.S. security, . . . near-term, mid-term, and long-term, would be the possibility of a terrorist organization obtaining a nuclear weapon."[1] According to Harvard University political scientist Graham Allison, who has emerged as the nation's leading voice of concern about nuclear terrorism, there is "better than a 50 percent chance that terrorists will detonate a nuclear bomb in the United States within ten years"[2] (that is, by 2014). A national commission on weapons of mass destruction concluded that there is a "better than even chance that terrorists will use biological or nuclear weapons within five years"[3] (that is, by 2013). The CIA's top analyst on terrorist use of weapons of mass destruction, Rolf Mowatt-Larssen, writes that now, in 2011, "we cannot exclude the possibility of nuclear terrorism. It is not tomor-

---

1 "U.S. President Barack Obama Warns of Nuclear Terrorism," BBC, April 12, 2010.

2 Graham Allison and Joanne J. Myers, *Nuclear Terrorism: The Ultimate Preventable Catastrophe*, New York: Times Books, 2004.

3 Bob Graham, Jim Talent, Graham Allison, Robin Cleveland, Steve Rademaker, Tim Roemer, Wendy Sherman, Henry Sokolski, and Rich Verma, *World at Risk: The Report of the Commission on the Prevention of WMD Proliferation and Terrorism*, New York: Vintage Books, 2008, p. xv.

row's threat; it is with us here today."[4] As former CIA Director Michael Hayden has warned: "Al Qaeda is the CIA's top nuclear concern."[5]

These are not the rantings of cranks with signs warning that the end is nigh; they are crafted public statements by some of America's highest-ranking, most highly respected, most thoughtful people. And yet al Qaeda, with no known nuclear capability, has ascended to the level of a virtual nuclear power, one that the CIA apparently ranks ahead of Iran, whose suspected nuclear-weapons ambitions are backed up by a large contingent of nuclear scientists and an extensive network of nuclear facilities; ahead even of North Korea, which we know possesses nuclear weapons. Al Qaeda has become the world's first terrorist nuclear power without, insofar as we know, possessing a single nuclear weapon.

## A Psychological Triumph

There is very little concrete evidence upon which to base such dire forecasts. The threat of nuclear terror floats far above the world of known facts. How has al Qaeda managed to pull off this stunning feat of psychological legerdemain?

The dramatic impact of terrorism provides part of the answer. What distinguishes terrorism from other modes of armed conflict is the separation between the actual targets of terrorist violence and the targets of the psychological terror. Because terrorists cannot hope to defeat their foes in open battle, they deliberately aim spectacular attacks at vulnerable civilian targets, hoping to create an atmosphere of terror, which will induce the public audience to exaggerate the terrorists' strength and dissuade governments from pursuing policies opposed by the terrorists because of the perceived price. Terrorist attacks also create political crises, provoking overreaction and compelling governments to divert vast resources to security in order to maintain public confidence

---

[4] Rolf Mowatt-Larssen, *Islam and the Bomb: Religious Justification For and Against Nuclear Weapons*, Cambridge, Mass.: Harvard Kennedy School, January 2011, p. 9.

[5] "CIA Chief: Al-Qaida Is Top Nuclear Concern," Associated Press, September 16, 2008.

that they will be protected, even while knowing that absolute security is not possible. This is the very essence of terrorism. And it often works.

The trajectory of contemporary terrorist violence also has contributed to fears of mass destruction. Nearly 40 years ago, observing that most terrorist violence was symbolic, I suggested that "terrorists want a lot of people watching, not a lot of people dead." Those earlier generations of terrorists seemed to worry that wanton slaughter would alienate their constituencies and be counterproductive to their political causes. While today's terrorists still argue about the appropriate level of violence, self-imposed constraints clearly have eroded.

As war has become less lethal, terrorism has become more lethal. In the decades since World War II, military power has moved away from the industrial-scale slaughter of total war, placing greater emphasis on reducing collateral casualties to a minimum. The development of increasingly precise weapons has facilitated this effort. Domestic genocide, particularly the targeting of specific ethnic groups, has been the exception to this rule. Meanwhile, contemporary terrorists have moved in the opposite direction, toward large-scale indiscriminate violence, escalating from small, mostly symbolic bombings involving few casualties in the 1970s to truck bombs aimed at killing hundreds in the 1980s and 1990s to the attacks on September 11, 2001, that killed thousands. From this cataclysmic event, it was easy to extrapolate the idea of Osama bin Laden using a nuclear weapon if he had one.

The 9/11 terrorist attacks fundamentally altered perceptions of plausibility. With box cutters and mace, terrorists turned commercial airliners into guided missiles that brought down skyscrapers. People feared that al Qaeda would try to launch more 9/11-scale attacks if it could, or perhaps even more-ambitious attacks. Terrorist scenarios that had been deemed far-fetched before 9/11 became operative presumptions after 9/11. In this environment, no terrorist scheme could be dismissed. Nuclear terrorism ascended to a clear and present danger.

There are vast differences between chemical, biological, radiological, and nuclear weapons. In its final report, written before 9/11, the National Commission on Terrorism chose wisely to avoid the collective term "weapons of mass destruction."[6] Aggregating such weapons confuses a low threshold for occurrence—terrorists already have employed chemical and biological weapons, with modest results—with a high potential for theoretical casualties, thereby exaggerating both probability and likely consequences. Realistically, only biological and nuclear weapons have a capacity for true mass destruction. And nuclear weapons differ from biological weapons in that biological weapons also may be used to kill just a few, as the anthrax letters did. It is hard to imagine a minor nuclear attack.

## Ambitions, Not Capabilities

The assessment of al Qaeda's nuclear capabilities is based on very little information. Everyone agrees that al Qaeda's leaders have nuclear ambitions. Osama bin Laden, while still in Sudan, may have had contact with some radicalized U.S. scientists, and according to one informant, he tried unsuccessfully to acquire nuclear material. Al Qaeda's abundant cash and lack of nuclear expertise made it an easy mark for scams. There were several reported cases in which al Qaeda thought it was purchasing nuclear-weapons components or fissile material but instead got nothing more than low-grade fuel, car parts, or other useless junk.

Al Qaeda's interest in nuclear weapons continued in Afghanistan. Documents left behind as al Qaeda retreated from the country included descriptions and even crude diagrams of nuclear weapons. A knowledgeable physicist who was asked to examine these documents said that they did not indicate the knowledge necessary to make a nuclear bomb.[7] Although their author apparently understood general theory, a bomb based on the diagrams would not have produced a nuclear explosion. Given time, the terrorists might figure out how to

[6] L. Paul Bremer and Maurice Sonnenberg, *Countering the Changing Threat of International Terrorism*, Report from the National Commission on Terrorism, June 2000. As of June 16, 2011: http://www.fas.org/irp/threat/commission.html

[7] David Albright, "Al Qaeda's Nuclear Program: Through the Window of Seized Documents," *Policy Forum Online*, Nautilus Institute, Special Forum, No. 47, November 6, 2002.

assemble a nuclear device, but they would still need fissile material and technical expertise. (Some more conspiracy-minded individuals opined that al Qaeda deliberately left behind amateurish documents to deceive authorities about just how far along its nuclear-weapons efforts actually were.)

Aware of al Qaeda's technological shortcomings, bin Laden sought the assistance of two sympathetic Pakistani scientists with whom he spoke in Afghanistan shortly before 9/11. They told him that he could not build a device with the material he had. It is not clear how much more technical advice the Pakistanis were able to provide, however. Within a few months, al Qaeda was on the run, and the two scientists ended up in the custody of Pakistani authorities.

How serious were al Qaeda's nuclear efforts? When he was interrogated about al Qaeda's nuclear weapons, Khalid Sheikh Muhammad, al Qaeda's chief operational planner and the architect of the 9/11 attacks, reportedly said the efforts never went beyond downloads from the Internet. There is also a report that some in al Qaeda doubted that the organization could develop true weapons of mass destruction but nonetheless agreed to continue to use the term because it would give the movement psychological influence.[8] Subsequent public reports suggest that even as al Qaeda's senior leaders were on the run, with their capability to launch large-scale terrorist operations diminished, efforts to acquire nuclear weapons continued. But despite numerous sensational "revelations" of the organization's purported nuclear arsenal, there is no evidence that al Qaeda has ever acquired a stolen nuclear weapon or the fissile material to make one.

Beyond these very few known facts about al Qaeda's nuclear aspirations, everything else falls into the realm of surmise and speculation, and there is plenty of that. Intelligence analysts, frustrated at the lack of hard evidence but unwilling to risk being accused of another "failure of imagination" as they were after 9/11, cannot exclude the possibility that al Qaeda or some other terrorist group might acquire nuclear weapons in the future. The absence of concrete evidence, therefore,

---

[8] Peter L. Bergen, *An Oral History of al Qaeda's Leader, The Osama bin Laden I Know*, New York: Free Press, 2006, p. 343.

does not diminish perceptions of the threat. Instead, the inability to say "it will never happen" can be taken a step further as confirmation that nuclear terrorism is inevitable—to use the now famous phrase, a matter of "not if, but when."

## Absence of Warning

Paradoxically, the absence of evidence heightens the threat. Many believe that the public will have no warning of an impending terrorist nuclear attack. Although there are two diametrically opposite views of U.S. intelligence capabilities, both fuel nuclear terror.

The first view of U.S. intelligence credits it with omniscience but remains suspicious that Washington is deliberately withholding information to avoid causing public panic. This is a popular theme among conspiracy-driven books and articles on nuclear terrorism that promise to tell the reader what the government will not. In fact, since 9/11, the government has routinely passed threat information—some of it vague, even dubious—on to the public. Yet it is also true that since the 1970s, there have been scores of undisclosed nuclear threats to American cities. The Federal Bureau of Investigation (FBI) and U.S. Department of Energy nuclear emergency teams have mobilized to conduct secret searches. All of the threats were found to be apparent hoaxes, but none of this was published at the time. As an example, just after 9/11, the CIA received intelligence from a source appropriately code-named Dragonfire that terrorists had planted a nuclear device in New York City. The federal government initiated a search without informing the public or local authorities. Nothing was found—the source was mistaken. But given this record, it is conceivable that if the government received credible intelligence that al Qaeda or another terrorist organization had acquired a nuclear weapon, it might not reveal it.

The second view of U.S. intelligence is that it cannot be depended upon to provide advance warning of a terrorist nuclear attack. We will know it only when we see the bright yellow flash. After all, American intelligence officials were surprised to discover in 1991 that Iraq had come closer to developing nuclear weapons than they had imagined.

U.S. intelligence did not foresee 9/11. It failed to predict the testing of nuclear weapons by India, Pakistan, and North Korea, but prior to the Iraq War, it reported with confidence that Iraq had weapons of mass destruction when it had none. While this recitation is not entirely fair to the intelligence community, perceived past failures of intelligence do not inspire confidence.

Lacking hard evidence that terrorists have nuclear weapons or material, intelligence analysts instead pore over al Qaeda's public statements for warnings or other clues about its interest in nuclear weapons. They argue over whether *fatwas*—religious rulings—authorizing al Qaeda to kill millions should be interpreted as the obligatory warning required by Islamic concepts of warfare. They debate whether terrorists might be deterred from acquiring or using nuclear weapons. These are legitimate lines of inquiry, but they also reify the threat, treating a hypothetical—al Qaeda's or any terrorists' possession of nuclear weapons—as if it were a concrete fact, or at least an inevitable development.

## Hypothetical Possession, Vivid Consequences

"A 10-kiloton nuclear bomb detonated in Times Square in New York City could kill a million people," noted U.S. Secretary of State Hillary Clinton in April 2010. "Beyond the human cost, a nuclear terrorist attack would also touch off a tsunami of social and economic consequences across our country."[9]

How is it that such alarming pronouncements come to be made, despite their having the predictable effect of exaggerating the likely threat and contributing to the terrorists' goal of instilling fear? As indicated by the string of quotes that began this essay, Secretary Clinton is hardly alone in her use of such pronouncements, nor was her calm discussion of nuclear arms reduction, nuclear proliferation, and nuclear

---

[9] Hillary Rodham Clinton, "Remarks on Nuclear Proliferation at the University of Louisville," Louisville, Ky., April 9, 2010.

terrorism intended to be scare-mongering. There is no sinister intent. The dilemma, generically, is that to do sensible things, political leaders must engage the enthusiasm of those charged with the "doing," must obtain funding from the fickle U.S. Congress, and must convey a sense of seriousness. They cannot do that with meek speeches that say, "Well, of course, there is also the outside threat of nuclear terrorism, but I don't lose sleep over it." They would lose credibility and stature, would be seen by the citizenry as soft and ineffectual, and could demoralize those on whom the country is counting to hunt down the bad guys.

Nuclear terrorism has thus created its own orthodoxy. Regardless of its likelihood, to question it is to risk being seen as soft. Whatever the intention, the formula of presentation is by now well established: A hypothetical event produces historically confirmable consequences.

Historically confirmable they are. While the threat of nuclear terrorism remains clouded in uncertainty, we know a lot about nuclear explosions. Hiroshima and Nagasaki and decades of Cold War experience in calculating nuclear blast effects provide detailed and vivid accounts of them. Analysts generally assume—the operative word—that a terrorist bomb would produce a 10-kiloton nuclear explosion, but it is by no means certain that terrorists could build a nuclear device with that yield. A crude nuclear device might just as likely be in the tenths-of-a-kiloton range, but that does not convey the desired dramatic impact.

A typical article on the subject begins with the assumption that terrorists can steal or build a nuclear bomb, then proceeds to describe in detail the intense fireball, blinding flash, diameter of immediate destruction, widespread firestorms, charred flesh, death by radiation, and social chaos that would emanate from a 10-kiloton blast at city center. That the underlying premise of terrorist possession of a nuclear weapon is hypothetical gets lost among the "reality" of its effects. Likewise, hypothetical worst-case scenarios generate consequences that outweigh their low probabilities of occurrence. With a presumed million fatalities—the equivalent of 300 9/11s—how low must the odds be to be considered acceptable?

## Spinning Nuclear Fantasies

Al Qaeda's efforts to obtain nuclear weapons were accompanied by an active communications effort, which implied that the terrorist group was further along in its quest than it actually was. This public communications campaign intensified as al Qaeda's central leadership came under increasing pressure after 9/11.

In earlier interviews, Osama bin Laden, when asked by reporters about weapons of mass destruction, coyly responded that their acquisition was a religious duty. But in an interview with a Pakistani reporter in November 2001, as the Taliban and al Qaeda were being bombed by American warplanes, bin Laden and Ayman al-Zawahiri were said to have claimed that al Qaeda had chemical and nuclear weapons, although the reporting of this interview raised doubts. The first published version had bin Laden saying only that al Qaeda would survive even if the United States used chemical or nuclear weapons against it; subsequent versions of the same interview introduced the claim that al Qaeda itself already had such weapons.

Neither of al Qaeda's two top leaders made many public mentions of nuclear weapons after the 2001 interview, but in 2002, an al Qaeda spokesman posted a message on the Internet claiming that because the United States was responsible for the deaths of millions of Muslims, al Qaeda, in accordance with Islamic law, had the right to kill 4 million Americans.[10] This was amended by a fugitive Saudi cleric who issued a religious ruling in 2003 authorizing al Qaeda to kill 10 million Americans.[11]

Exactly why al Qaeda elicited the two statements about killing millions of Americans is not known, but the statements excited the organization's followers and alarmed analysts in the United States. The two communications prompted a lively discourse among jihad-

---

[10] Suleiman Abu Gheith, "In the Shadow of the Lances," June 2002, translation at *MEMRI Special Dispatch Series,* No. 388, June 12, 2002.

[11] Nasser bin Hamad al-Fahd, "Risalah Fi'istjkhdam 'asliha Al-Dammar Al-Shamil Did Al-Kuffar," May 2003. For a discussion, see Reuven Paz, "Global Jihad and WMD; Between Martyrdom and Mass Destruction," *Current Trends in Islamist Ideology,* Vol. 2, September 2005.

ists about al Qaeda's nuclear posture and strategy, as if its possession of nuclear weapons were real. Analysts in the United States, meanwhile, interpreted the two statements as providing the necessary warning before attack required by the Islamic code of warfare. They noted that the only way al Qaeda could achieve this magnitude of casualties was with nuclear, or possibly biological, weapons. It was seen as a signal.

Al Qaeda's communications campaign was not a centrally directed effort but, rather, a distributed project, a new phenomenon made possible by the Internet. A chorus of online jihadists carried on the campaign, issuing threats and adding lurid landscapes of fireballs and mushroom clouds over Manhattan and Washington. These were the fantasies of the powerless, vicarious participation in al Qaeda's terrorist campaign. Psychologically satisfying to their jihadist authors, they kept Western government officials on edge.

Al Qaeda's nuclear threats also commanded the attention of the U.S. news media, which, for reasons of commercial competition, has veered off into the realm of shock and entertainment in the years since 9/11. Editorial constraint yielded to drama. The line between news and fiction blurred. Here was the stuff of suspense novels, with Western civilization hanging in the balance. What could make for a better story? Nuclear terror acquired a life of its own.

## Driven by Our Imagination

A receptive audience, Americans have been avid consumers of nuclear terror. It has revived Cold War anxieties. It has resonated with many who anticipate the Apocalypse, confirming their belief that we are in the end times. It has acted as a condenser for the nation's broader apprehensions about economic and political decline, its fear of being taken over by alien cultures.

"Nuclear terrorism" and "nuclear terror" have different domains of meaning. Terrorism is action, and it can include the mere threat of action, which can produce its own powerful effect. Terror is the effect. Nuclear terrorism today is about the possibility that terrorists will acquire and use nuclear weapons, while nuclear terror is about the *anticipation* of that event. Nuclear terrorism is driven by terrorist capabilities. Nuclear terror is driven by our imagination.

The history of nuclear terrorism can be quickly summarized: It hasn't happened. Terrorists are not known to have acquired and certainly have not used nuclear weapons—although many would hasten to add "yet." But nuclear terror is real, and it has become deeply embedded in our popular culture and policymaking circles.

This fear does not come free. It adds another layer to our already considerable national anxieties. It fans xenophobia. It corrodes our commitment to liberty. It demands obedience to an orthodoxy that cannot be challenged without provoking accusations of being dangerously naïve or soft on terrorism. And yet, paradoxically, despite the widespread apprehension that within a very few years, terrorists armed with nuclear weapons will destroy an American city, we still live in those cities. Even those who warn us that nuclear terrorists are coming have not removed themselves from danger.

The government did not invent the bogeyman of nuclear terrorism to compensate for the end of the Cold War or to perpetuate the military-industrial complex, as some cynics suggest. Concern about the nuclear threat is genuine, and we should be grateful for the many officials who work tirelessly to prevent nuclear attacks from any quarter. But we should also be careful about how we allow the threat of nuclear terrorism to be harnessed to serve a variety of purposes, some of which are more useful than others.

## Harnessing the Power

Ironically, the dramatic excesses in nuclear terror since 9/11 may be having some useful consequences. The heightened fears, in particular, may be stimulating nations worldwide, and factions within them, to become more unified in demanding and supporting stringent measures to control nuclear weapons and related materials, thereby strengthening the consensus that nuclear weapons are to be avoided, reduced, and maybe even eliminated. The merits of any policy moving toward the latter are debatable for complex reasons, including verification, but such a broad consensus would bode well for mankind.

The threat of nuclear terrorism might be used to counter arguments that because some countries have nuclear weapons, others are entitled to have them. While countries may disagree on how to treat suspected *proliferators*, they can more easily agree that too many nuclear-weapons programs, too many nuclear-weapons designers, too much fissile material, and too many nuclear weapons increase the threat of nuclear *terrorism* and therefore agree that proliferation itself is dangerous to the entire world. Even if proliferation cannot be stopped in its tracks, raising concerns about nuclear terrorism can open space for cooperation on increasing controls, reducing access, and implementing other positive measures.

The threat of nuclear terrorism can just as easily be employed to argue for nuclear disarmament on grounds that the elimination of existing nuclear arsenals, or at least significant reductions in them, will reduce the threat of loose nuclear weapons and therefore of nuclear terrorism. Whether the threat of nuclear terrorism is real, and whether preventing further proliferation or disarmament would reduce that threat, is not the issue. It is a matter of manipulating widespread perceptions for the greater good.

At the very least, the threat of nuclear terrorism, real or imagined, has led to increased security for nuclear weapons and fissile material. That must be counted as a positive development, however it comes about.

If nuclear terror has prompted measures that are good in and of themselves, that is to be applauded, but has it prevented nuclear terrorism? Some assert that by raising the alarm, Armageddon has been averted—at least, deferred—but this circular argument treats its own presumptions as fact. One cannot prove that the threat of nuclear terrorism has been exaggerated or, absent evidence, that it has ever existed. Nor can one prove that it is now more likely or less imminent.

So far, at least, nuclear terrorism has occurred only in novels. Nuclear terror, on the other hand, is a fact. What matters now is whether we are its victims or its masters.

## Related Reading

Ackerman, Gary A., Charles P. Blair, Jeffrey M. Bale, Victor Asal, and R. Karl Rethemeyer, *Anatomizing Radiological and Nuclear Non-State Adversaries: Identifying the Adversary.* College Park, Md.: National Consortium for the Study of Terrorism and Responses to Terrorism, 2009.

Bass-Golod, Gail V., and Brian Michael Jenkins, *A Review of Recent Trends in International Terrorism and Nuclear Incidents Abroad*, Santa Monica, Calif.: RAND Corporation, N-1979-SL, 1983. As of May 24, 2011: http://www.rand.org/pubs/notes/N1979.html

Daly, Sara A., John V. Parachini, and William Rosenau, *Aum Shinrikyo, Al Qaeda, and the Kinshasa Reactor: Implications of Three Case Studies for Combating Nuclear Terrorism*, Santa Monica, Calif.: RAND Corporation, DB-458-AF, 2005. As of May 24, 2011: http://www.rand.org/pubs/documented_briefings/DB458.html

Davis, Paul K., and Brian Michael Jenkins, *Deterrence and Influence in Counterterrorism: A Component in the War on Al Qaeda,* Santa Monica, Calif.: RAND Corporation, MR-1619-DARPA, 2002. As of May 24, 2011: http://www.rand.org/pubs/monograph_reports/MR1619.html

DeLeon, Peter, Bruce Hoffman, Konrad Kellen, and Brian Michael Jenkins, *The Threat of Nuclear Terrorism: A Reexamination*, Santa Monica, Calif.: RAND Corporation, N-2706, 1988. As of May 24, 2011: http://www.rand.org/pubs/notes/N2706.html

Jenkins, Brian Michael, "Georgia Dispute Derails Bid to Stop Nuke Terrorism," *Providence Journal*, October 6, 2008.

———, *The Likelihood of Nuclear Terrorism*, Santa Monica, Calif.: RAND Corporation, P-7119, 1985. As of May 24, 2011: http://www.rand.org/pubs/papers/P7119.html

———, "A Nuclear 9/11?" CNN.com, September 11, 2008.

———, "Nuclear Terror: How Real?" *Washington Times*, May 13, 2007.

———, *The Potential for Nuclear Terrorism*, Santa Monica, Calif.: RAND Corporation, P-5876, 1977. As of May 24, 2011: http://www.rand.org/pubs/papers/P5876.html

———, *Terrorism and the Nuclear Safeguards Issue*, Santa Monica, Calif.: RAND Corporation, P-5611, 1976. As of May 24, 2011: http://www.rand.org/pubs/papers/P5611.html

———, *Will Terrorists Go Nuclear?* Los Angeles, Calif.: Crescent Publications, 1975.

———, *Will Terrorists Go Nuclear?* Amherst, N.Y.: Prometheus Books, 2008.

———, "Will Terrorists Go Nuclear?" *United Press International*, September 11, 2008.

Levi, Michael A., *On Nuclear Terrorism,* Cambridge, Mass.: Harvard University Press, 2007.

Masse, Todd M., *Nuclear Jihad: A Clear and Present Danger?* Dulles, Va.: Potomac Books, Inc., 2011.

Meade, Charles, and Roger C. Molander, *Considering the Effects of a Catastrophic Terrorist Attack*, Santa Monica, Calif.: RAND Corporation, TR-391-CTRMP, 2006. As of May 24, 2011: http://www.rand.org/pubs/technical_reports/TR391.html

Mueller, John, *Atomic Obsession: Nuclear Alarmism from Hiroshima to Al-Qaeda,* Cambridge: Oxford University Press, 2009.

Parachini, John V., David E. Mosher, John C. Baker, Keith Crane, Michael S. Chase, and Michael Daugherty, *Diversion of Nuclear, Biological, and Chemical Weapons Expertise from the Former Soviet Union: Understanding an Evolving Problem*, Santa Monica, Calif.: RAND Corporation, DB-457-DOE, 2005. As of May 24, 2011: http://www.rand.org/pubs/documented_briefings/DB457.html

Wenger, Andreas, and Alex Wilner, *Deterring Terrorism: Theory and Practice,* forthcoming.

Wilner, Alexandre S., "Deterring the Undeterrable: Coercion, Denial, and Delegitimization in Counterterrorism" *Journal of Strategic Studies*, Vol. 34, No. 1, pp. 3–37.

# PART THREE
# Torn Between Physical Battles and Moral Conflicts

Since 9/11, Americans have had to think differently about warfare, suggests Christopher Paul. The reality of asymmetric warfare, the fact that not all conflicts can be won through force alone, and the contested superiority of American values—these factors have grown more prominent over the past decade, making warfare today more complex, at least tactically, than it was before.

"Whether fighting terrorists or insurgents," writes Paul, a combat-oriented approach alone "does nothing to address the underlying popular motives that lead to terrorist or insurgent movements in the first place." In such conflicts, America needs to extend "both a closed fist and an open palm . . . winning the physical battles while addressing the grievances" of those who support America's adversaries.

Paul's research on the past 30 years of counterinsurgency worldwide has shown that successful counterinsurgents employ a host of beneficial practices simultaneously to prevail in physical combat while also gaining the moral support of the broader population. It is a difficult challenge to pursue both "security *and* economy," he underscores, "legitimacy *and* pacification."

Both sides of the coin are dicey propositions.

As for military pacification, there is no guarantee that America will succeed in its core strategic mission of eliminating terrorist havens around the world, notes Kim Cragin. "Just because havens might represent viable targets does not mean that we can eliminate them."

The current policy debate has produced no viable alternative strategy for defeating terrorist havens in a complicit state other than "to invade it, occupy it, gain control over its people and territory, and establish a democratic government." The long odds and high costs of doing so have presented America with a moral conflict that hits close to home, Cragin asserts.

"It makes me pause and think about the many soldiers, sailors, Marines, and airmen who have repeatedly risked their lives in the fight against al Qaeda. It makes me reflect on the times that I have looked into the eyes of soldiers as they pled with me to accept their needs to resign because they 'couldn't watch buddies get killed again for no reason.'" The alternative strategy Cragin proposes for countering terrorist havens is isolation, not elimination.

As for addressing grievances and gaining legitimacy, there is no guarantee that America will succeed in this mission, either, notes Todd Helmus. Al Qaeda has taken advantage of U.S. political bravado. "Extremists have also been delighted with the ill-conceived anti-Muslim attitudes and behavior of a small minority of Americans." Moreover, "pernicious anti-Muslim statements have become increasingly common in American blogs, talk radio shows, and even political rallies. These bigoted sentiments play right into al Qaeda's hands."

Recent improvements on many of these fronts "have not gone unnoticed" in the Muslim world, Helmus acknowledges. Still, in moving forward, "the United States must continue to be driven by the better angels of its nature. . . . We must disprove the claims of our enemies by our deeds."

CHAPTER EIGHT

# Winning Every Battle but Losing the War Against Terrorists and Insurgents

*Christopher Paul*

*Christopher Paul is a RAND social scientist whose areas of focus include strategic communication, counterinsurgencies, and information operations.*

The 9/11 terrorist attacks have changed how Americans think about war by emphasizing different challenges. My research on counterterrorism and counterinsurgency has highlighted three of these newly salient (though not new) challenges and what they mean for America as it attempts to win the war, not just the battles, against terrorist foes. In brief, these newly salient changes have required America to use both a closed fist and an open palm at the same time, winning the physical battles while addressing the grievances of, and reaching shared understandings with, those who support America's adversaries.

*The first challenge is that the future surely includes asymmetric warfare.* The past decade has shown the potential effectiveness of asymmetric attacks against a stronger foe. Suicide attacks, improvised explosive devices, sniping, the use of human shields, and the intermingling of combatants and non-combatants are all tactics that minimize the ability of a conventionally stronger opponent to bring that strength to bear.

Asymmetric warfare is here to stay. As long as the U.S. military remains the world's dominant conventional fighting force, its adversaries will sensibly seek to avoid confronting it head on. Meanwhile, U.S. national interests are served by being engaged abroad and by conducting stability, peacekeeping, and nation-building operations, which send forces beyond U.S. borders and expose them to asymmetric threats.

Moreover, a growing number of current and likely future threats will involve non-state adversaries. Such foes, lacking any national resources, have little choice but to engage in asymmetric warfare if they wish to fight.

*The second challenge is that not all conflicts can be won through force alone.* As U.S. General David Petraeus articulated in his 2010 counterinsurgency guidance to the International Security Assistance Force deployed to Afghanistan: "We can't win without fighting, but we also cannot kill or capture our way to victory."

This is particularly true of conflicts involving non-state actors or other foes seeking asymmetric advantages. Whether fighting terrorists or insurgents, a strictly "kinetic" (combat-oriented) solution does nothing to address the underlying popular motives that lead to terrorist or insurgent movements in the first place. These unaddressed motives remain as impetuses for new terrorist or insurgent recruits and for those who support them.

Superior equipment and martial skill might allow the United States to win all the battles, but trying to prevail in this way could also condemn it to losing the war. Strictly military defeat of terrorist or insurgent groups can have the effect of multiplying their ranks, as still others take up the cause or seek retaliation. The application of force alone does nothing to address their reasons for fighting.

*The third challenge is that the rightness of American values and the wrongness of extremism are not self-evident.* When the United States seeks to do good in the world or to promote values that it assumes are universal—or, at the very least, would be judged as positive by any reasonable external standard—it should not be surprised when others do not see things the same way. "Freedom" and "democracy" have different connotations in other parts of the world and are not always viewed in positive terms. Sometimes, they raise suspicion of ulterior motives, and a history of promoting democracy in one place while supporting a dictator in another sends a contradictory message to populations and foreign governments.

Conversely, the violent acts of extremists are not universally reviled by everyone everywhere. Acts of commission by extremists (the 9/11 attacks, for example) are compared to acts of omission on the part of the United States (sanctions leading to starvation in Iraq) and are judged to be morally similar. Religiously inspired governance and justice offered by extremists can compare favorably with corrupt governance and capricious justice offered by secular dictatorships.

Even where particular acts of violence are held to be reprehensible, the causes that precipitate those acts can be considered noble or honorable. Morally, we do not live in a black-and-white world, and there is considerable nuance in how others view us and our adversaries.

Combining these last two challenges puts things in even sharper focus: The United States cannot kill or capture its way to victory, because the underlying motives for hostility remain, and those motives will not be sufficiently diminished by self-evident righteousness.

## Diminishing Support

So what must be done? The rediscovered wisdom on counterinsurgency, dating back to a 1963 RAND report by David Galula on attempts by colonial France to pacify Algeria,[1] if not before, holds that popular support is the insurgents' center of gravity. My research on terrorism and insurgency reaffirms this conclusion, with an added nuance. Although support is indeed the critical center of gravity, support can come either from the local population, from a state, from an external diaspora, or from international sales of insurgent-controlled natural resources, such as diamonds, timber, or narcotics.[2] While popular support is qualitatively the "best" kind of support, the nuance is that *all* sources of support matter. Even when counterinsurgent forces succeed in cutting off an insurgency from popular support from within the country, if signif-

[1] David Galula, *Pacification in Algeria*, 1956–1958, Santa Monica, Calif.: RAND Corporation, MG-478-1-ARPA/RC, 2006 (originally published in 1963). As of May 24, 2011: http://www.rand.org/pubs/monographs/MG478-1.html

[2] Daniel Byman, Peter Chalk, Bruce Hoffman, William Rosenau, and David Brannan, *Trends in Outside Support for Insurgent Movements*, Santa Monica, Calif.: RAND Corporation, MR-1405-OTI, 2001. As of May 27, 2011: http://www.rand.org/pubs/monograph_reports/MR1405.html

icant support continues to flow from extranational sources, insurgents will likely be able to continue operating.[3]

The nuance also extends to separating "support" into two forms: expressed support and tangible support. Those who express support for a terrorist or insurgent group are glad that the group exists, endorse (at least some of) the group's actions, and would like to see (at least some of) the group's stated objectives met. This kind of support is often measured by public opinion polls. Tangible support, in contrast, is that which supplies tangible needs, such as personnel, materiel, funding, sanctuary, and intelligence.[4]

Whatever removes popular motivations for expressed support (by lessening sympathy, decreasing shared identity, or addressing grievances, for example) will decrease the pool of possible recruits, threaten the legitimacy of extremist movements, diminish their moral force, and decrease the amount of tangible support provided by local populations. Tangible support that is not provided by local populations must be reduced by different means. Smuggling routes can be interdicted, bank accounts seized, and state sponsors pressured by the international community.

When both expressed and tangible support are substantially diminished, the threat from a terrorist or insurgent group will also be substantially diminished, but not necessarily eliminated. Even if recruiting is more difficult, it will never be impossible. Even if materiel is scarce, there will always be enough for simple (but still dangerous) attacks. Even if amnesty is offered or other forms of disarmament, demobilization, and reintegration are pursued, some diehard fighters will decline the overtures. Any terrorist or insurgent group is likely to leave a residual hard core of individuals who must be captured or killed in order to finally eliminate the residual threat. But the diminution of support would make it harder for these villains to hide, and diminished

[3] See Christopher Paul, Colin P. Clarke, and Beth Grill, "Victory Has a Thousand Fathers: Evidence of Effective Approaches to Counterinsurgency, 1978–2008," *Small Wars Journal*, January, 2011.

[4] See Christopher Paul, "As a Fish Swims in the Sea: Relationships Between Factors Contributing to Support for Terrorist or Insurgent Groups," *Studies in Conflict & Terrorism*, Vol. 33, No. 6, 2010, pp. 388–410.

sympathy among previously supportive (or just tolerant) populations would make helpful intelligence tips more likely. The trick, then, is to apprehend or otherwise deal with this residual threat without creating a chain of events that renews motivations for participation and support.

## Winning the Battles *and* the Wars

Without question, the United States must continue to win the battles, maintaining its martial excellence and eliminating adversary forces when presented with opportunities to do so. However, unless it also reduces the support given to irregular and asymmetric opponents, the United States will have to win the same battles over and over again.

Winning the war will not come about because of the inherent "goodness" of the United States or the intrinsic "evil" of the extremists, but because of the actual actions and utterances of the United States. Often described as "the battle for hearts and minds" (a phrase that I personally dislike because it is unnecessarily militant and oppositional), the kinds of things described under that rubric are critical to decreasing feelings of support. In general, these efforts aim to inform, influence, and persuade the populations of concern (and the people of the world more generally) that the United States is and intends to be a force for good, that it shares more values and views than they might think, and that the extremist worldview is a threat to everyone's way of life.

The challenge is to fight the physical battles and to engage the broader population at the same time. But we know that it works, and we know how it is done. For U.S. military units in Iraq, success often arose when commanders were able to simultaneously employ multiple mutually reinforcing lines of operation. My research on the past 30 years of insurgency worldwide also shows that successful counterinsurgents always employ a host of beneficial practices simultaneously.[5] Security

---

[5] See Christopher Paul, Colin P. Clarke, and Beth Grill, *Victory Has a Thousand Fathers: Sources of Success in Counterinsurgency*, Santa Monica, Calif.: RAND Corporation, MG-964-OSD, 2010. As of May 27, 2011: http://www.rand.org/pubs/monographs/MG964.html

*and* economy. Democracy *and* development. Legitimacy *and* pacification. Influence *and* direct action.

What we have learned since 9/11 has several implications for the future. Here are three.

First, prevailing over terrorists and insurgents requires simultaneously combating armed adversaries and diminishing their support, which is tricky. It is far better to apply a healthy dose of prophylactic action to prevent situations that are likely to create terrorists or insurgents from emerging in the first place. This may mean getting involved in international development, peace support, security cooperation, public diplomacy, and other forms of global engagement earlier if doing so can reduce the need to deploy combat forces later.

Second, a counterinsurgency strategy should identify all sources of support and seek to diminish them simultaneously or as they arise. Tangible support can come from the population, but it can also come from foreign sources or from both the population and foreign sources. A counterinsurgency strategy should not assume that the population is the only source of support or the only center of gravity.

Third, even with strong preventive measures, there will be unforeseen contingencies, and the United States must be ready for them. Readiness means whole-of-government approaches to counterterrorism and counterinsurgency, including greater resources for civilian agencies, as well as military forces equipped and prepared to win all the battles, but to do so in ways that avoid losing the wars.

## Related Reading

Connable, Ben, and Martin C. Libicki, *How Insurgencies End*, Santa Monica, Calif.: RAND Corporation, MG-965-MCIA, 2010. As of May 24, 2011:
http://www.rand.org/pubs/monographs/MG965.html

Davis, Paul K., Kim Cragin, eds., *Social Science for Counterterrorism: Putting the Pieces Together*, Santa Monica, Calif.: RAND Corporation, MG-849-OSD, 2009. As of May 24, 2011:
http://www.rand.org/pubs/monographs/MG849.html

Gompert, David C., John Gordon IV, Adam Grissom, David R. Frelinger, Seth G. Jones, Martin C. Libicki, Edward O'Connell, Brooke Stearns Lawson, and Robert E. Hunter, *War by Other Means—Building Complete and Balanced Capabilities for Counterinsurgency: RAND Counterinsurgency Study—Final Report*, Santa Monica, Calif.: RAND Corporation, MG-595/2-OSD, 2008. As of May 24, 2011: http://www.rand.org/pubs/monographs/MG595z2.html

Headquarters, U.S. Department of the Army, and Headquarters, U.S. Marine Corps, *Counterinsurgency Field Manual*, Field Manual 3-24/Marine Corps Warfighting Publication 3-33.5, Chicago, Ill.: University of Chicago Press, 2007.

Long, Austin, *On "Other War": Lessons from Five Decades of RAND Counterinsurgency Research*, Santa Monica, Calif.: RAND Corporation, MG-482-OSD, 2006. As of May 24, 2011: http://www.rand.org/pubs/monographs/MG482.html

Metz, Steven, and Raymond Millen, *Insurgency and Counterinsurgency in the 21st Century: Reconceptualizing Threat and Response*, Carlisle, Pa.: Strategic Studies Institute, U.S. Army War College, 2004.

Paul, Christopher, *Strategic Communication: Origins, Concepts, and Current Debates*, Westport, Conn.: Praeger Security International, 2011.

Paul, Christopher, Colin P. Clarke, and Beth Grill, *Victory Has a Thousand Fathers: Detailed Counterinsurgency Case Studies*, Santa Monica, Calif.: RAND Corporation, MG-964/1-OSD, 2010. As of May 24, 2011: http://www.rand.org/pubs/monographs/MG964z1.html

———, *Victory Has a Thousand Fathers: Sources of Success in Counterinsurgency*, Santa Monica, Calif.: RAND Corporation, MG-964-OSD, 2010. As of May 24, 2011: http://www.rand.org/pubs/monographs/MG964.html

Ross, Michael L., "What Do We Know About Natural Resources and Civil War?" *Journal of Peace Research*, Vol. 41, No. 3, May 2004, pp. 337–356.

CHAPTER NINE

# The Strategic Dilemma of Terrorist Havens Calls for Their Isolation, Not Elimination

*Kim Cragin*

*A RAND senior cultural historian focusing on terrorism, Kim Cragin spent three months of 2008 on the staff of U.S. General David Petraeus in Iraq. She has conducted extensive fieldwork in such areas as the West Bank and Gaza Strip, Lebanon, Colombia, northwest China, Sri Lanka, Pakistan, Indonesia, and the southern Philippines.*

It was spring 2009 when I met some friends for dinner and drinks in Washington, D.C. The topic of conversation was the Obama administration's new strategy for Afghanistan and Pakistan. Two of us were terrorism researchers, a third worked for the U.S. Department of Defense, and the fourth had just spent a year in Pakistan. We were asking ourselves, Is this fight truly about Afghanistan, or is it about al Qaeda? And if it is about al Qaeda, does the new White House strategy go far enough?

At the time, similar conversations were occurring in the hallways of U.S. government buildings, on editorial pages, and in academic forums. The White House had just released its new policy toward Afghanistan and Pakistan. This policy placed al Qaeda's safe havens at the center of the new U.S. strategy against al Qaeda, stating, "The core goal of the U.S. must be to disrupt, dismantle, and defeat al Qaeda and its safe havens in Pakistan, and to prevent their return to Pakistan or Afghanistan."[1]

[1] The White House, *White Paper of the Interagency Policy Group's Report on U.S. Policy Toward Afghanistan and Pakistan*, Washington, D.C., March 2009. As of May 27, 2011:

Cynics could argue that the White House chose this goal to justify an increased focus on Afghanistan and Pakistan. But it is also possible that those writing the policy were convinced that the best way forward against al Qaeda was to eliminate its sanctuaries in South Asia. Either way, placing safe havens at the core of the U.S. strategy against al Qaeda created a strategic dilemma: Right now, the only way to disrupt, dismantle, and defeat such havens in a complicit state is to invade it, occupy it, gain control over its people and territory, and establish a democratic government. No viable alternative exists in the current debate. And while this solution may be appropriate for Afghanistan, it seems unreasonable and unlikely to work against other havens. As one U.S. defense official said to me, "Of course, the fight is really in Pakistan, but we can *do something* about Afghanistan."

I find this rationale to be deeply discouraging. I would like to think that I am pragmatic enough to accept that just because al Qaeda has vulnerabilities does not necessarily mean that the United States has been able to exploit them effectively. Yet it makes me pause and think about the many soldiers, sailors, Marines, and airmen who have repeatedly risked their lives in the fight against al Qaeda. It makes me reflect on the times that I have looked into the eyes of soldiers as they pled with me to accept their needs to resign because they "couldn't watch buddies get killed again for no reason."

Meanwhile, the expense of sustaining military forces in Afghanistan amounts to an estimated $1.1 million per soldier per year, during a time of economic hardship. If havens are indeed viable targets, then the United States needs to identify a better way of dismantling them in the future, after it withdraws from Afghanistan. After all, it has become clear that even if the United States and NATO manage to stabilize population centers in Afghanistan, the Taliban will continue to offer havens to al Qaeda in the more remote border areas of Afghanistan and Pakistan. So beyond the immediate effort in Afghanistan and the recent successful operation against Osama bin Laden in Pakistan, it is likely that al Qaeda havens will continue to exist in Pakistan, Yemen,

---

http://www.whitehouse.gov/assets/documents/afghanistan_pakistan_white_paper_final.pdf

Somalia, and other locations for the foreseeable future. The alternative strategy that I propose against these al Qaeda havens would be isolation, rather than elimination.

## The Ongoing Challenge

The strategic dilemma about how best to tackle al Qaeda's havens now being confronted by the Obama administration has confounded its predecessors of both political parties for more than two decades. The Clinton administration faced similar challenges after the 1998 U.S. embassy bombings in Kenya and Tanzania that were planned and executed by al Qaeda operatives, killing 231 people, including 12 Americans. As they are today, the policy instruments available in 1998 were limited because the U.S. security apparatus had been built around threats from states, not non-state actors. Two weeks after the embassy bombings, President Clinton approved retaliatory attacks with cruise missiles against al Qaeda training camps and facilities in Afghanistan and Sudan.

In announcing these attacks, Clinton gave the following justification: "The United States does not take this action lightly. Afghanistan and Sudan have been warned for years to stop harboring and supporting these terrorist groups. But countries that persistently host terrorists have no right to be safe havens."

The immediate response to an attack against U.S. interests overseas during the Clinton administration was to attack al Qaeda's training camps and facilities and to punish those countries harboring its operatives. Why? In its in-depth exploration of the 1998 embassy bombings and the U.S. response, the 9/11 Commission concluded that senior U.S. officials had considered multiple avenues of response, both in the immediate aftermath and for a continuing campaign against al Qaeda, but most were discarded because of a perceived lack of feasibility or for the sake of general caution. Thus, arguably, cruise missiles were chosen as the immediate response because a cruise-missile attack was something the United States could do. Subsequently, the United

States adopted a plan to capture Osama bin Laden, a plan that was given only a 15-percent chance of success by the CIA at the time.

The George W. Bush administration faced similar problems. In her April 2004 testimony before the 9/11 Commission, National Security Adviser Condoleezza Rice described the dilemma as she saw it prior to the 9/11 attacks: "Integrating our counterterrorism and regional strategies was the most difficult and the most important aspect of the new strategy to get right. Al Qaeda was both a client of and a patron to the Taliban, which in turn was supported by Pakistan. Those relationships provided Al Qaeda with a powerful umbrella of protection and we had to sever that."[2]

Safe havens again. Administration after administration—Republican and Democrat—have returned to terrorist havens as the key hurdle to overcome.[3]

And yet experts have disagreed about whether havens are important for al Qaeda to pursue its global strategy. In his October 2009 testimony before the Armed Services Committee of the U.S. House of Representatives, noted terrorism expert Paul Pillar argued that an expansion of military efforts in Afghanistan was unwarranted: "It is not apparent to me how a move of al Qaeda or parts of it from one side of the Durand line to the other would substantially affect the threat the group poses to U.S. interests. Any such threat would be no less from Waziristan [Pakistan] than it would be from Nuristan [Afghanistan]."[4]

Just one month earlier, in his own testimony before the Senate Foreign Relations Committee, John Nagl, president of the Center for a New American Security, argued the exact opposite: "American efforts now focus on Pakistan as a launching pad for transnational terrorists and insurgents fighting in Afghanistan. But the problem runs both

---

[2] "Testimony of Condoleezza Rice Before 9/11 Commission," *New York Times*, April 8, 2004. As of May 27, 2011: http://www.nytimes.com/2004/04/08/politics/08RICE-TEXT.html?pagewanted=3

[3] Other countries—including Israel, India, Colombia, Spain, and the United Kingdom—have faced similar challenges in dealing with havens.

[4] Paul Pillar, Hearing Before the Full Committee of the Committee on Armed Services, House of Representatives, Washington, D.C., October 14, 2009. As of July 5, 2011: http://www.gpo.gov/fdsys/pkg/CHRG-111hhrg56004/html/CHRG-111hhrg56004.htm

ways: a failed Afghanistan would become a base from which the Taliban and al Qaeda militants could work to further destabilize the surrounding region."[5]

Last year, several RAND researchers examined al Qaeda strategic documents to discern the views of al Qaeda's senior leaders on the need for physical sanctuary or lack thereof. (We felt that, strangely, in this debate among experts, the perspectives of al Qaeda's senior leaders had been absent.) Included among the al Qaeda documents studied was a work by one of its forefathers, Abdullah Azzam; statements released in the mid-1990s by al Qaeda's Advisory and Reform Committee in London; and more recent treaties by strategist Abu Musab al-Suri, as well as guidance provided by second-in-command Ayman al-Zawahiri to Al Qaeda in the Arabian Peninsula released in its magazine, *Sada al-Malahim*. Our analysis led us to conclude that while al Qaeda could learn to function without a physical sanctuary, its leaders believe that they need a safe location from which to pursue a global strategy, and thus they behave accordingly.

On the basis of this review, it is reasonable to conclude that consecutive White House administrations have focused on havens. Unfortunately, the only solution before us still is to invade and occupy a sanctuary, which brings us back to the familiar strategic dilemma: Just because havens might represent viable targets does not mean that we can eliminate them.

## Time for a Different Approach

This lesson has been one of the hardest for me to learn in my ten years at RAND: It is sometimes tempting to believe that describing a complex problem is enough, without taking on the burden of offering a solution. However, I have come to reject that mindset over the years, as I repeatedly have encountered military commanders, charged with *doing*

---

5 John Nagl, Hearing Before the Committee on Foreign Relations, United States Senate, September 16, 2009. As of July 5, 2011: http://www.gpo.gov/fdsys/pkg/CHRG-111shrg55538/html/CHRG-111shrg55538.htm

*something*, who are looking for new ideas. This reality struck me again when I briefed an audience of a half-dozen three- and two-star generals last year in the Pentagon—a 20-minute briefing that was followed by 90 minutes of discussion about how to tackle certain aspects of al Qaeda. The study sponsor was thrilled that the research generated such an extensive discussion on U.S. counterterrorism policy. And, indeed, it still ranks as one of my most rewarding interactions with senior policymakers. Still, I left the meeting somewhat dissatisfied with myself, as a statement from one of the attendees rang in my head: "Okay, I buy that [this topic] represents a key vulnerability of al Qaeda, but I don't see what we can do about it."

Fair enough. So following along these lines, as al Qaeda havens along the Afghanistan-Pakistan border (and elsewhere) appear to merit the strategic focus of consecutive White House administrations, what can be done about them?

First, the time has come for us to admit that we will not be able to dismantle al Qaeda havens in Pakistan or other parts of the world in the short-to-medium term. Beyond drone strikes and isolated raids, there is very little the United States can do militarily outside of Afghanistan and Iraq. Unless the United States attempts to put large numbers of "boots on the ground" in multiple havens, only host nations themselves can dismantle those within their borders. Only Pakistan will be able to eliminate havens within its borders; Yemeni forces, in their country; and so on. Currently, U.S. forces are providing the Pakistani Army and Frontier Scouts with counterinsurgency training. But the relationship between Pakistan and the United States is unpredictable at best. Moreover, past experience, as in Colombia and the Philippines, would suggest that this type of training tends to take years, if not decades, to bring about change, even under the best of circumstances.

Second, we should recognize that drone strikes are not effective in dismantling physical havens. Drone strikes clearly make al Qaeda feel less safe and can even disrupt operational planning, but that is not the same thing as rendering a haven unusable. In a February 2010 article entitled "The Year of the Drone," published by the New America Foundation, Peter Bergen and Katherine Tiedemann argued that while drone strikes have wrought havoc on al Qaeda morale, they have not discouraged new Western recruits from flocking to northwest Pakistan (between 100 and 150 recruits made the journey in 2009 alone), nor have they discouraged cross-border attacks into Afghanistan.[6] In fact, the most we can hope for from the drone strikes is the elimination of key individuals and perhaps others fleeing to separate areas of, for example, Pakistan outside the reach of U.S. drones, areas such as Karachi. This perhaps represents a shifting of havens, but not their complete dismantling.

Third, I would suggest that the United States government, rather than focusing on eliminating al Qaeda havens, should design a strategy to isolate them. Make it difficult for al Qaeda leaders to communicate with their followers and sympathizers outside Pakistan. Pressure allies to arrest al Qaeda couriers (rather than just monitor them) as they move back and forth between Pakistan and other parts of the Muslim world. Place financiers on blacklists so that their own personal funds are rendered inaccessible and they are unable to travel. Expand the roles and presence of other U.S. government agencies—such as the Federal Bureau of Investigation, Drug Enforcement Administration, and Immigration and Customs Enforcement—in places where the traditional counterterrorism forces are unwelcome or controversial. Many al Qaeda members and sympathizers are involved in other illegal activities, such as kidnapping, human smuggling, or narcotics, that these law enforcement agencies are better able to exploit. This basic approach would not preclude drone strikes and other military efforts but would shift the focus from primarily military activities to include law enforcement as well. Of course, some of this is already being done on a small scale, but these law enforcement efforts tend to be treated as less-important and supporting activities, rather than priorities.

This alternative approach of isolating rather than eliminating havens would not be foolproof. It would have limitations. Fundamentally, it would abdicate any U.S. effort to occupy and control territory

---

[6] Peter Bergen and Katherine Tiedemann, "The Year of the Drone: An Analysis of U.S. Drone Strikes in Pakistan, 2004–2010," Washington, D.C.: New America Foundation, February 24, 2010. As of June 18, 2011: http://counterterrorism.newamerica.net/sites/newamerica.net/files/policydocs/bergentiedemann2.pdf

that al Qaeda might use as a haven. And because the United States would not control territory, its intelligence collection and ability to respond to potential attacks might be reduced. But is the United States really going to invade and occupy al Qaeda havens again in the near future? If not, isolation would appear to be the only way forward.

## Related Reading

Bergen, Peter, *The Osama bin Laden I Know: An Oral History of al-Qaeda's Leader*, New York: Free Press, 2008.

Byman, Daniel, *Deadly Connections: States that Sponsor Terrorism,* New York: Cambridge University Press, 2005.

Davis, Paul K., and Kim Cragin, eds., *Social Science for Counterterrorism: Putting the Pieces Together*, Santa Monica, Calif.: RAND Corporation, MG-849-OSD, 2009. As of May 24, 2011:
http://www.rand.org/pubs/monographs/MG849.html

Lia, Brynjar, *Architect of Global Jihad: The Life of al-Qaeda Strategist Abu Mus'ab al-Suri*, New York: Columbia University Press, 2008.

Tucker, David, *Skirmishes at the Edge of Empire: The United States and International Terrorism*, Washington, D.C.: Praeger, 1997.

For more al Qaeda strategy documents and other primary source material, see the West Point Combating Terrorism Center's website, http://www.ctc.usma.edu/categories/publications/harmony-documents

CHAPTER TEN

# Our Own Behavior Can Be Our Weakest Link—or Our Strongest Weapon

*Todd C. Helmus*

*A RAND clinical psychologist specializing in terrorism and insurgency, Todd Helmus has served as an adviser in counterinsurgency strategy to U.S. military commands in Iraq and Afghanistan.*

They say a picture is worth a thousand words. In 2004, news broke that U.S. prison guards at Abu Ghraib were abusing and humiliating Iraqi detainees. The pictures that vividly told the tale went viral. U.S. casualties in Iraq immediately spiked. For years afterward, the Abu Ghraib scandal was a rallying cry and motivator for foreign fighters who streamed into Iraq to martyr themselves in suicide operations.

Ten years after 9/11, al Qaeda is still an inspiration and proverbial destination for far too many terrorist recruits. A mainstay motivation of extremists has been the perception that the West, and the United States in particular, is at war with Islam. Al Qaeda specializes in using U.S. policies and actions, such as the abuse at Abu Ghraib, to make its case. The United States can remove much of this ammunition by being ever sure that its conduct meets the highest standards and reflects its noblest ideals.

Individuals join the jihadist cause for a multitude of reasons. The excitement of a clandestine and militant life, the recognition and fame that are all too rare in a moribund home life, the social bonds of friends and peer groups that act as a propellant for action, the lure of an afterlife that gives more than this life offers—all of these factors motivate terrorist recruits at one level or another.

But the most common motivation appears to be the perception that a war is being waged against Islam. This is at the heart of the jihadist narrative. The worldwide Muslim community, or *ummah*, is purportedly under attack by infidel and Western powers. Al Qaeda thus calls Muslims to Islam's defense; and to al Qaeda, it is not a defense of rhetoric but a defense of armed action. To the vast majority in the Muslim world, the appeal has no resonance. For a hapless and dangerous few, however, it is all too compelling.

## Showing, Not Just Telling

In propounding this war against Islam, America's enemies often use outright lies and fabrications. But they also know that their message is most convincing and most motivating when it draws on real events. It is one thing to tell recruits that America is at war against Islam and another to show actions that seem to prove it. Consequently, al Qaeda has been eager to take advantage of America's own policies, pronouncements, and behaviors to crudely and unfairly illustrate its message.

Even before Abu Ghraib, the U.S. war in Iraq was seen by some as reason enough to join the militant cause. In the eyes of jihadists, the Iraq War was unjustified and constituted a foreign and infidel occupation of Muslim lands. In this sense, the motivations of foreign fighters entering Afghanistan in the 1980s were little different from those of fighters coming into Iraq. Simplistic and false accusations that the United States entered Iraq to "take its oil," the real fact that the United States ignored many pleas of the international community before invading, and the U.S. failure to establish security following a successful invasion were further used to tarnish the image of the United States.

Al Qaeda uses U.S. operations in Afghanistan to similar effect. In the wake of 9/11, it was necessary to attack al Qaeda at its Afghan base, and establishing stability and security remains an imperative. But the war drags on. The rallying cry of an infidel occupation of a Muslim land is a common jihadist refrain that not only motivates foreign fighters and some Taliban but also galvanizes recruits in the West. This refrain was the argument used by Nidal Hassan, the sole suspect in the shootings at Fort Hood, Texas, whose PowerPoint presentation to classmates at the Uniformed Services University of the Health Sciences spoke of the war against Islam, the U.S. campaign in Afghanistan, and a Muslim's duty.

The very nature of the Iraqi and Afghan conflicts, wars fought among civilian populations in an urban landscape, has inevitably led to accusations of abuse. Insurgents and terrorists in both countries have foresworn distinguishing uniforms and shamefully hide behind civilian populations. U.S. forces by no means intend civilian casualties and go to great lengths to avoid them. But when mistakes happen, al Qaeda pounces. The isolated and worst-case scenarios are used to allege a systematic campaign of abuse. Aiding these arguments, at times, has been a slow and feeble response that in the cases of Abu Ghraib, the deaths of civilians in Haditha, and other civilian casualty events has failed to quickly investigate accusations and admit wrongdoing. Perceptions of abuse of Muslims were what motivated Carlos Bledsoe, the convicted attacker at the military recruiting center in Little Rock, Arkansas, who asserted that he was defending Muslims against U.S. military actions in the Middle East.

## Shooting Oneself

After 9/11, the United States took an aggressive stance toward detainees, who were given the status of unlawful combatants and denied prisoner-of-war status, held without trial or habeas corpus, and, in some instances, subjected to torture and harsh interrogations. The status of detainees in a war without uniforms or nation-states has fostered a complex debate with merit on both sides. But complex and nuanced arguments are quickly stripped away once they are put through al Qaeda's propaganda machine, and our detainee policy, as evidenced by opinion polling conducted by World Public Opinion and others, has been perceived as still more evidence of U.S. ill intent.

Al Qaeda has also taken advantage of U.S. political statements and bravado. President George W. Bush announced the "Global War on Terror" nine days after 9/11 to define the conflict of a new age. However, the term *terror* is much broader than *terrorism*, so instead of al Qaeda being the focus of the fight, America's adversary became a conflated mix of groups that often had little association with the militant network that had attacked it. Many Muslims throughout the Middle East and beyond came to perceive the sweeping term as evidence that Islam itself was the target of American retribution. While certainly not intended, U.S. pronouncements sowed confusion where none was required.

Extremists have also been delighted with the ill-conceived anti-Muslim attitudes and behavior of a small minority of Americans. The most extreme example was the "Burn a Koran Day" that was to be hosted by a small fringe church in Gainesville, Florida. The event was canceled at the last minute at the urging of U.S. officials, but its mere planning and publicity stirred resentment against the United States and spurred violent threats on extremist websites.

In a similar fashion, many Americans found controversy in plans to build a Muslim community center and mosque near Ground Zero. Pernicious anti-Muslim statements have become increasingly common in American blogs, talk radio shows, and even political rallies. These bigoted sentiments play right into al Qaeda's hands, not only seeming to prove its point to future extremists but also alienating an American Muslim population that has played a critical role in clamping down on homegrown extremism.

Today's media environment works to al Qaeda's advantage. At one time, extremists had to compile elaborate videotapes depicting horrendous abuse of Muslims and then pass them from one person to the next. For many, the message was compelling, but the dissemination process was slow. Today, however, an image from a distant battlefield or small-town America can be instantly uploaded on YouTube and onto jihadist websites. The videos go viral and pass from one to many. Jihadist websites wrap these images in a narrative that argues that such pictures are not the exception but the rule. The images are accompanied by stirring music and grand exhortations. The number of jihadist websites and chat rooms has proliferated into the thousands, reaching Arabic and English speakers alike.

A visit to the Internet has become a first and potent step in the slow process of radicalization. The discontent and angry can while away hours in front of the computer screen. They easily ignore other web content that fails to justify their extremist views. They gain reassurance that their perceptions, albeit misplaced, were right all along. They interact in chat rooms and web forums where their anger is reinforced in a virtual social environment. Many view attack videos that illustrate an exciting and seemingly efficacious outlet for discontent. Surely, most would-be jihadists just sit and stew in a virtual world. Others, though, gain real-world connections and find solace among peer groups that cycle the anger further and galvanize the will.

The radicalization of a few is fed by the resentment of many. Opinion polls throughout the Middle East testify to the unfavorable perceptions of U.S. policy. Large majorities hold negative opinions of the United States and believe that it intends to weaken and divide Islam. Majorities claim that the West does not respect Islam. Most in the Arab and Muslim world do not support terrorist attacks, but their overall opinions matter because they expand the pool of individuals who may be open and receptive to one of al Qaeda's core messages. The overall perceptions create a social environment in which hatred seems acceptable and may in fact be reinforced in everyday conversation. For some who become so imbued with this ideology, the only step left is to act.

## Counter Resentment, Counter Terrorism

In this war of ideas, perceptions can matter more than reality. It is surely fair to argue that American service personnel entering Iraq did so to remove a brutal dictator who killed hundreds of thousands of Muslims, and it is certainly true that U.S. involvement in Afghanistan is intended to bring stability to a war-torn country and that the attitudes of a few Americans do not represent those of a tolerant American society. Unfortunately, the enemy will spin what it can spin, and it will take the worst example and hold it up as emblematic of all that America is. These perceptions—and the ease with which they can be disseminated—demand that we as Americans continuously hold ourselves to an ever-higher standard.

Nothing written here is new to the United States. The inflaming consequences of the war in Iraq are well understood. The U.S. military, previously unaccustomed to dealing with insurgents and civilians alike, has significantly reduced civilian casualties, markedly improved oversight of detention facilities, and gained a hard-earned focus on interacting with and protecting civilian populations. The torture of detainees is no longer permitted, and those at the Guantanamo Bay detention camp are increasingly having their days in court. The United States increasingly speaks of engagement with the Muslim world and favors international cooperation over unilateral action. These changes have not gone unnoticed.

In moving forward, the United States must continue to be driven by the better angels of its nature. U.S. leaders must think through their policies in advance to understand how they will affect and be interpreted by the Muslim world. The United States must not seek short-term gains at the expense of long-term counterterrorism and humanitarian objectives. The U.S. military must continually orient the force to execute missions successfully and carefully in civilian-populated areas. Mistakes and misconduct, when uncovered, should be followed by public acknowledgment and assurances that they will not be repeated. Politicians and civic and religious leaders must work together to quash the seeds of hatred that threaten American unity.

Abu Ghraib was a stain on the character of America, but it is a stain that must not be repeated or be allowed to define who America is to the world. A picture is indeed worth a thousand words, but we as a nation can determine how the future pictures will be drawn.

Misperceptions about U.S. intent will still abound and will feed the hardened hearts of a few. The war against al Qaeda will be long. The fight in Afghanistan will continue. Terrorist networks will be hunted. Mistakes will surely be made, and well-meaning actions and statements will be deliberately misrepresented. We as Americans must be prepared for this. But more to the point, we must disprove the claims of our enemies by our deeds.

## Related Reading

Coll, Steve, *Ghost Wars: The Secret History of the CIA, Afghanistan, and bin Laden, from the Soviet Invasion to September 10, 2001*, New York: Penguin Press, February 2004.

Davis, Paul K., Kim Cragin, eds., *Social Science for Counterterrorism: Putting the Pieces Together*, Santa Monica, Calif.: RAND Corporation, MG-849-OSD, 2009. As of May 24, 2011:
http://www.rand.org/pubs/monographs/MG849.html

Helmus, Todd C., Christopher Paul, and Russell W. Glenn, *Enlisting Madison Avenue: The Marketing Approach to Earning Popular Support in Theaters of Operation*, Santa Monica, Calif.: RAND Corporation, MG-607-JFCOM, 2007. As of May 24, 2011:
http://www.rand.org/pubs/monographs/MG607.html

Horgan, John, *The Psychology of Terrorism*, New York: Routledge, 2005.

Kilcullen, David, *The Accidental Guerrilla: Fighting Small Wars in the Midst of a Big One*, Oxford: Oxford University Press, March 2009.

Sageman, Marc, *Leaderless Jihad: Terror Networks in the Twenty-First Century*, Philadelphia, Pa.: University of Pennsylvania Press, January 2008.

Wright, Lawrence, *The Looming Tower: Al-Qaeda and the Road to 9/11*, New York: Vintage Books, September 2007.

# PART FOUR
# Driven by Unreasonable Demands

In reaction to 9/11, Americans not only demanded immediate action to prevent future terrorist attacks but also kindled expectations that "could not be satisfied by any realistic homeland security policies," writes Brian Jackson in his history of post-9/11 America. "Fear drove action," he concludes, "and political rhetoric frequently stoked rather than cooled the flames of urgency."

Even as time passed, the sense of urgency persisted, as if the nation had been overrun by day traders. "Just as the chief executive officers of public companies complain that their investors demand short-term performance improvements on a quarterly basis, making it impossible to build and invest for stronger long-term performance, the nation—or at least policymakers—acted like a 'short-term security investor,' wanting better homeland security, right now."

The flurry of unrealistic demands could be neither fulfilled nor sustained. The great haste drove undue losses in dollars, in organizational effectiveness, in accountability, and even in the fight against terrorism. "For terrorists who in part were trying to frighten the country into squandering resources out of sheer panic, the country gave them good reason to feel encouraged."

It would be far more effective for the country to pursue security measures that promise long-term, rather than urgent and short-term, benefits. Airline security is a case in point, according to Jack Riley: "The focus of these efforts should be on what works best and at least cost, but the traveling public remains doubtful that the focus has been put in the proper place."

Today, the security regime for the flying public dictates the same rigid procedures for all 700 million passengers who board planes each year in the United States. "That we have not developed a reasonable way to reduce that inspection workload is perhaps the biggest missed opportunity of the past decade," Riley asserts. "It is imperative that we use the next decade to develop smarter, more sustainable, and more practical solutions to air passenger security." Riley offers a solution, based on a trusted traveler program for flights originating in the United States combined with continued strict procedures for travelers coming to America.

Exaggerated reactions to the terrorist threat are understandable—to a point. "The relative obscurity of the terrorist enemy . . . might help explain why Americans fear it well beyond the real danger," writes Gregory Treverton. At the same time, politicians "abet those fears by seeming to imply that *any* terrorist attack can be prevented."

Intelligence analysts can abet those fears, too. "Intelligence naturally focuses on new threats and worst cases and so contributes to the fear without really meaning to do so. It is time for U.S. intelligence agencies to rethink not only how they do their business but also how they can do that business while adding less to the climate of terror that gives the terrorists a win."

America's intelligence community has made commendable progress since 9/11, says Treverton, but "politicians and intelligence officials alike need to do a better job of speaking plainly to the American people. Terrorism, as long as terrorists do not obtain nuclear or highly destructive biological weapons, is more a nuisance than a mortal threat."

CHAPTER ELEVEN

# Don't Let Short-Term Urgency Undermine a Long-Term Security Strategy

*Brian A. Jackson*

*Brian Jackson is a RAND senior physical scientist whose research focuses on homeland security and terrorism preparedness. His areas of examination have included terrorist organizational behavior, use of technology by terrorist organizations, threat analysis, emergency preparedness, incident management in large-scale emergency-response operations, the equipment and technology needs of emergency responders, and the design of preparedness exercises.*

The 9/11 terrorist attacks were neither the first incidents of terrorism on U.S. soil nor even the first attacks on the World Trade Center or the Pentagon. In 1993, the World Trade Center was attacked with a vehicle bomb by Ramzi Yousef, a prolific terrorist subsequently connected to major plots aimed at the aviation system. Decades before, in 1972, the Weather Underground, a domestic group, successfully detonated a bomb inside the Pentagon. Previous domestic terrorist attacks had also resulted in numerous casualties; one such attack, the bombing of the Murrah Federal Building in Oklahoma City in 1995, killed and injured hundreds of people. But the nature and aftermath of the 9/11 attacks set them apart from all that had come before and produced a reaction among policymakers and the public that was different from those following previous terrorist attacks.

Even though there had been some discussion before the attacks about the potential for large-scale terrorist attacks—including attacks with unconventional weapons—the reality of such an attack shocked the nation. Probably because its consequences differed so dramatically from public expectations, 9/11 caused an abrupt shift in national assumptions about terrorism. In some policy circles, terrorism—and terrorism from al Qaeda in particular—began to be spoken of as an existential threat to the United States, a language that had seldom been used since the Cold War, when the Soviet nuclear arsenal truly did threaten the survival of the nation. In the policy debate that followed 9/11, the threat of terrorism was also linked closely to the issue of weapons of mass destruction, with statements that it was not a matter of *if*, but *when* a terrorist attack using unconventional weapons would be waged on the homeland. These shifts meant that the *prevention* of future terrorist attacks, rather than responses to them after they occurred, became much more central in most thinking about security policy.[1] An understandable sense of urgency dominated the policy debate.

The perception of a driving need to act catalyzed major shifts in policy. Government agencies were reorganized. Money moved. Large security-technology projects, including major systems to secure the borders or detect nuclear threats from abroad, were moved quickly from conception to implementation in an effort to ramp up protection for the nation as soon as possible. Fear drove action, and political rhetoric frequently stoked rather than cooled the flames of urgency. Even as time passed after 9/11, this sense of urgency persisted, both motivating a continued focus on security and shaping the initiatives intended to protect the U.S. homeland. Just as the chief executive officers of public companies complain that their investors demand short-term performance improvements on a quarterly basis, making it impossible to build and invest for stronger long-term performance, the nation—or at least policymakers—acted like a "short-term security investor," wanting better homeland security, right now.

In political and policy debate, what "better" meant was also undefined. At times, an extremely high standard was set. The national

[1] This privileging of prevention was similarly embedded in the 2002 *National Security Strategy of the United States of America* and its articulation of a willingness to act preemptively against apparently threatening actors or states.

discourse on terrorism and security for years after the attacks was rife with internal contradictions. Though many reasonable policymakers acknowledged publicly that perfect security was unattainable, over the same period, the rhetoric of some members of Congress—from both sides of the aisle—included statements that implied exactly the opposite. Such statements ranged from general exhortations that *no* future attacks on the country would be tolerated or that the "overriding goal" of policy was attack prevention[2] to more overtly political arguments that even the acknowledgment by members of the executive branch that future attacks might occur was evidence of the failure of current security policies.[3] Some members of the executive branch themselves defined the goal of security efforts as prevention of all future attacks.[4]

---

[2] For example, Christopher Cox commented in the March 25, 2004, hearing on the Terrorist Screening Center, "Our overriding objective, after all, is to prevent terrorist attacks." Comments from the Director of the New York State Office of Public Security in that same hearing ("Our first and foremost responsibility is to do everything humanly possible to prevent another terrorist attack . . .") echo this sentiment from another level of government. Statements that the nation had to ensure that terrorist attacks could not "ever happen again" echoed those made in legislative branch discussion and debate immediately after the September 11 attacks themselves. Policy arguments regarding specific security measures also argued implicitly that the goal was prevention of every attack, and very high costs would be justified if the measure prevented even a single attack (remarks by Scott McInnis on profiling in a November 7, 2001, congressional debate; John Kerry's remarks on rail security, December 14, 2001).

[3] For example, in a speech made in the Senate on May 23, 2002, Daniel Akaka said, "Vice President Cheney's remark on Sunday that the question is 'when' not 'if' a terrorist will attack the United States suggests that the administration has not met its most basic mission of homeland security and the war on terrorism." John Bolton's remark on a Fox News program on September 22, 2010, that comments made by President Obama to Bob Woodward that the United States could absorb a terrorist attack suggested that the President "doesn't care about Americans dying" shows that such statements can come from either side of the political spectrum and they continue to occur nearly a decade after the 2001 attacks.

[4] For example, in December 2001 post-testimony discussion before the Senate Judiciary Committee, Attorney General John Ashcroft said, "Very frankly, we have set as a priority the prevention of additional terrorist attacks, and we don't ever want anything like September 11th again to visit the United States, on our own soil with innocent victims. And we hope to improve our performance regularly, by making whatever changes we can to upgrade our ability to detect and to prevent terrorism, to disrupt it and to make it difficult, in fact impossible." Internal accounts from the executive branch, including those in Bob Woodward's books and Ron Suskind's *The One Percent Doctrine,* similarly indicated that policymakers

Others were more circumspect and were sometimes punished for it.[5] Assumptions about terrorist access to and ability to use weapons of mass destruction permeated this debate, since if large-scale unconventional attacks are indeed viewed as inevitable and potentially posing an existential threat to the country, intellectual self-consistency essentially requires pursuing policies that seek to prevent every attempted attack.[6]

Legislative mandates for 100-percent screening of air cargo and inbound shipping containers reflect the logic that missing even one attack attempt is unacceptable. This perceived need to preempt attacks was part of a political and policy discussion that shifted terrorism from being viewed as "a law enforcement matter" to being viewed as a national security or military one. Another example of this change could be clearly seen in domestic counterterrorism, where the FBI was pressured to "move faster" to *disrupt* nascent terrorist plots—rather than extending surveillance to gather stronger evidence or more intelligence on the participants and their associates—to minimize the chance that plotters might slip away and be able to carry out their attacks.

---

defined the goal of policy as assuring that there would be no additional attacks in the U.S. homeland.

[5] For example, in remarks on March 26, 2007, TSA Administrator Kip Hawley forthrightly acknowledged that 100-percent security was impossible. But when Department of Homeland Security Secretary Michael Chertoff stated in a media interview that the federal government could not be responsible for preventing every attack on mass transportation systems and needed to focus its efforts on preventing large-scale terrorism, members of Congress called his statements "appalling" and demanded an apology (for example, Senator Charles Schumer said on July 14, 2005, "I hope and pray that Secretary Chertoff misspoke, because every one of our citizens on transit—whether in the air, on the water, on rail, or on the road—is our responsibility to keep safe and prevent terrorism from afflicting them").

[6] See remarks by Senators Judd Gregg, Ted Stevens, and Richard Lugar, March 7, 2003, on terrorism and unconventional weapons. Senator Lugar said, "Terrorists, when armed with weapons of mass destruction, are in a position to create what philosophers would call existential events for countries. . . . One terrorist attack with a weapon of mass destruction has the potential to create such dislocations in the economy of a country that recovery could take decades. This existential threat from terrorism is a new condition for the world that requires changes in our policy priorities. All nations do not understand this with the same precision that the United States and our leadership does. All nations have not been attacked in the same manner we have been."

Though never promoting "prevention at all costs," the policy calculus certainly shifted toward "prevention at a much higher acceptable cost" than had been the case in previous years. On the national balance sheet for counterterrorism, the costs—in dollars, in American's civil liberties, in the perception of the United States by other nations and their citizens, and in other tangible and intangible factors—added up. U.S. policy choices had opportunity costs as well, as the paths chosen closed off other options and strategies that might have been pursued. For example, part of the price paid for the FBI's shift to more rapidly disrupting nascent plots was—for various reasons—an inability to successfully prosecute some of the alleged terrorists in the criminal justice system.[7] Other costs of counterterrorism and security became perceived as more acceptable: more intrusiveness in security searches at airports, less accessibility of public sites that might be attractive targets of attack, and so on. Urgency in implementing policies often meant that efforts were not made to measure whether those polices were actually performing. When they were, performance measures for some programs that suggested that 100-percent effectiveness was not being achieved—or was not the goal—were deemphasized or discarded.

The insistence on immediate action, compounded by the demand for very high levels of detection and prevention, meant that expectations could not be satisfied by any realistic homeland security policies. For terrorists who in part were trying to frighten the country into squandering resources out of sheer panic, the country gave them good reason to feel encouraged. Looking forward, it would be far more practical—and, most importantly, far more effective—for the country and its policymakers to shift national priorities toward a strategy focused

---

[7] An analysis of terrorism prosecutions by the New York University Law School Center on Law and Security cited an "early practice of making high-profile arrests while prosecuting few terrorism charges," shifting over time to more terrorism-specific prosecutions (Center on Law and Security, New York University School of Law, *Terrorist Trial Report Card: September 11, 2001–September 11, 2009*, January 2010). The prioritization of prevention over prosecution was clearly acknowledged by then–Attorney General John Ashcroft: "Our fight against terrorism is not merely or primarily a criminal justice endeavor—it is defense of our nation and its citizens. We cannot wait for terrorists to strike to begin investigations and make arrests. The death tolls are too high, the consequences too great. We must prevent first, prosecute second" (testimony before the House Judiciary Committee, September 24, 2001).

on building sustainable counterterrorism and homeland security efforts for long-term, rather than urgent and necessarily short-term, security benefits.

## Demanding Perfection Makes Failure Inevitable

Perfection is always a worthy ideal to aspire to, and the sentiment expressed by many policymakers in private that they would endeavor to do everything they could to ensure that America is not attacked again "on their watch" is entirely understandable. But perfection is also an unrealistic yardstick by which to measure any real-world security program, and a relentless urgency for short-term improvement makes it even less realistic. Perfection as a goal, particularly in security, is also one that invariably pushes cost upward. It therefore should surprise no one that such internal U.S. government watchdogs as agency Inspectors General and the Government Accountability Office have since identified problems with performance and escalating costs in many security programs. In the later years of the decade following 9/11, flagship programs like SBInet—the virtual border fence intended to help control the southern border as a component of the Secure Border Initiative—were even abandoned after expenditures of large sums of taxpayer money.

The undercurrent of unrealistic expectations that has pulled on the homeland security debate over the years has meant that even failed terrorist attacks have been interpreted as failures of security, producing political criticism, recrimination, and the pressure to "do more." Umar Farouk Abdulmutallab (variously known as the Christmas, Detroit, and underwear bomber) failed in his attempted attack on an incoming airliner in 2009 in part because security measures that were already in place had pushed him to a type of operation that was more likely to fail—and provided the passengers near him the opportunity to stop him in the process. Rather than being viewed as something like a partial success for the security measures that were in place at the time, the incident was seen as a complete failure—exemplified and potentially magnified by the ridicule cast upon U.S. Secretary of Homeland Security Janet Napolitano's overly rosy comments immediately after the attack that "the system worked." The attempted attack also drove the subsequent deployment of costly new body scanners with the potential to detect similar devices concealed beneath clothing.

The contrast between the homeland security debate and other policy discussions is striking. Criminal incidents, particularly those resulting in multiple homicides, generally receive intense public and media attention. While preventing such incidents is certainly a goal of law enforcement, there is no expectation that all (or even many) of them can be prevented, at least at any price we are willing to pay. Yet the end goal of government action in both cases is the same: protecting American lives from the actions of individuals with the intent to injure or kill.

Framing the goal of counterterrorism as risk *elimination*—which is what perfect or near-perfect prevention would amount to—rather than risk *management* also undermines efforts to build a society that is resilient to terrorist threats. Some observers have drawn a contrast between the American and British responses to terrorist threats in recent years, with the longer experience with terrorism in the United Kingdom producing a general public and a private sector more able to readily "bounce back" after an attempted or successful attack.[8] Such resiliency has been highlighted as a critical part of national preparedness given that perfect prevention is impossible.[9] Although the assumed need to adopt specific new security measures after Abdulmutallab's attempted attack seems to suggest that the United States lacks this

---

[8] It has been suggested that this contrast is breaking down, however. For example, Jonathan Evans, head of MI5 (the British domestic intelligence agency) was quoted as saying, "In recent years we appear increasingly to have imported from the American media the assumption that terrorism is 100 percent preventable, and any incident that is not prevented is seen as a culpable government failure. This is a nonsensical way to consider terrorist risk and only plays into the hands of the terrorists themselves" (Gordon Corera, "MI5 Head Warns of Serious Risk of UK Terrorist Attack," September 16, 2010. As of June 17, 2011: http://www.bbc.co.uk/news/uk-11335412).

[9] See, for example, discussion in the Homeland Security Advisory Council *Report of the Critical Infrastructure Task Force*, January 2006, p. 4: "Strategies based on resilience accept that efforts to prevent attacks (reduce threats) and to defend against those attacks (reduce vulnerabilities), albeit necessary, will inevitably prove insufficient."

resilience, that failed attempt in some ways demonstrates the opposite: Beyond the fact that the public played a direct role in stopping the attack, opinion polling before and after the attempt also showed that it produced a relatively muted reaction and little fear—demonstrating a level of resiliency.[10,11]

In the past decade, the United States has faced an ongoing threat from terrorist attacks, but none of those attacks—successful or disrupted—has approached the scale of what happened on 9/11, likely thanks in part to the improvements in security that have been made since. But the experience since 2001 also suggests—with the substantial advantage of hindsight—that our willingness to bear significant extra costs to "do something, as quickly as possible" was not justified, at least as it applied across the board to all security efforts. The concern in the aftermath of 9/11 was that the next attack would be comparable to or worse than those attacks, that the terrorist threat had transformed in scale, scope, and sophistication. Indeed, there have been successful attacks since 2001, but fortunately, they have resembled the "pre-9/11" terrorist threat much more than the sketches of "post-9/11" terrorism as an existential threat to the country.

Many of the disrupted plots fall squarely in the historical tradition of pre-9/11 violent groups and individuals, whose limits in skill and capability often lead to either the terrorists being apprehended before they can act at all or, when they do act, their attacks falling well below their aspirations. Although the loss of lives in terrorist actions is always regrettable, the fact that post-9/11 terrorism is more consis-

---

[10] For example, Gallup polling showed only a minor change in expressed fear of terrorism in polls before and after the attack attempt (Gallup, "U.S. Fear of Terrorism Steady After Foiled Christmas Attack," January 12, 2010. Available at http://www.gallup.com/poll/125051/U.S.-Fear-Terrorism-Steady-After-Foiled-Christmas-Attack.aspx). At the same time, however, a significant portion of the public (42 percent of respondents) expressed the opinion that counterterrorism measures taken in the wake of the attempt did "not go far enough"; 42 percent believed the measures either were "about right" or went "too far," and the remainder had no opinion.

[11] In testimony before the Senate Judiciary Committee on January 20, 2010, Suzanne Spaulding used the case of Abdulmutallab and the post-attempt policy debate to make the very point that policy should aim for risk *management* rather than seeking to *eliminate* risk by preventing every attack.

tent with the history of terrorist group behavior than we thought it would be means that—as a nation likely to face terrorist threats from varied sources over the long term—we have both the opportunity and the requirement to approach security planning deliberately rather than hastily, with a focus on how the benefits of the security measures we implement will compare with the costs of building and maintaining them.

Has the way we make security choices changed since 2001? Although the post-9/11 urgency has dissipated to an extent, it persists in some areas, and it is fed by domestic political discourse in which demands to take dramatic action against terrorist threats have proven to be effective rhetorical tools. Not every unsuccessful attack has resulted in urgent changes in security measures or policies, but some prominent ones certainly have. The best example is the deployment of body scanners in response to Abdulmutallab's attempted attack, though questions were raised almost immediately about the cost, benefit, and sustainability of that security measure. In spite of the questions, the implementation of that policy is going forward. So in many ways, Americans—or at least many of the policymakers acting in their stead—still act like short-term security investors rather than taking a longer-term view, and by doing so, they are ceding advantages in the fight against terrorism and offering them to the terrorist adversaries.

## Sustainable Security Equals Success

Americans need to take advantage of the lessons that experience since 9/11 can teach—and the strengths of existing counterterrorism efforts—and set aside the sense of urgency, opting instead for counterterrorism policies that can be sustainable and effective over the long term. A more deliberate approach is needed to make prudent, effective security investments while reducing the costs of protecting ourselves. In an increasingly fiscally constrained environment, allowing urgency to provoke the dissipation of resources in projects that produce no security benefits or only a fraction of the benefit promised is a particularly unattractive strategy. This is not to say that there will never be a future terrorist attack where short-term, even knowingly unsustainable, action might be needed—for example, responding to an attack if concerns regarding large-scale terrorist use of unconventional weapons actually do come to pass—but those actions should be quickly reassessed, and the drive to act quickly should be replaced with deliberation as soon as possible.

Urgency has costs that go beyond the dollar costs of unsuccessful security efforts or canceled technology programs. Urgent demands for improvement have also led the U.S. Department of Homeland Security, an organization formed not that long ago, as time is measured for large and complex bureaucratic organizations, to reorganize itself multiple times since its creation in 2003, driven in part to respond to criticism about its performance and well-recognized problems integrating the many organizations that were merged when it was founded. A continued short-term focus should not be allowed to deny the department the opportunity to consolidate itself, address its internal challenges, build up its human resources, and configure itself to function effectively over the long term.

The long view is also needed to match security efforts to the type of threats posed by modern terrorist groups. Al Qaeda and other terrorist organizations change their behaviors and tactics, often in response to the security measures put in place. Thus, it is important to avoid hastily deploying security solutions that attackers can quickly learn how to sidestep or defeat.

Moreover, it is important to not inadvertently help those adversaries by the way security is discussed and its performance described in policy and political debates. An unintended consequence of the episodic demand for perfect security, as expressed in policy discussion and debates, has been ceding a rhetorical advantage to our terrorist adversaries. Since any attack—of any scale, whether successful or disrupted—is treated as a failure of security, the terrorists have learned to treat any attempted attack as a success. In discussions of Abdulmutallab's disrupted attack, Al Qaeda in the Arabian Peninsula argued that failure was success, because the U.S. reaction to the attempt produced political disruption and Americans "terrified themselves" even though the bomb did not even go off.[12] Taking a more balanced view of what we demand from security measures not only will help us deploy resources more effectively, it will also deny terrorist attackers the chance, with potentially only the stroke of a pen, to turn their failures into successes.

Urgent demands to "act now" also make it difficult to build security programs and strategies that can be monitored and evaluated for the benefits they deliver and that provide policymakers the information they need to adjust and improve them over time. To sustain homeland security in the long run, we must build measures and metrics that can tell us how we are doing and why systems are performing as they are, thereby making it possible, when the next attack—attempted or successful—occurs, to learn and adjust our strategies more smoothly in response to changes by our adversaries, rather than responding in a knee-jerk way.

## Time to Reduce the Pendulum Swings in Security Policy

Another major risk of security measures that are enacted in haste and under pressure is the fact that they ultimately defeat themselves—and not just because of their financial costs. Historians of domestic security in the United States frequently use the metaphor of a pendulum to describe how the perceptions of urgent threats that demand immediate responses can increase public willingness to pay for and tolerate more security. The initial pendulum "push" of fear enables more invasive, more intensive, and more costly security interventions because of the perceived need to act. For example, when concerns about international communism dominated early in the Cold War, public acceptance of domestic security activities increased. But when the fear dissipated, the pendulum swung back, and concerns about the way government agencies had been using their new powers and capabilities caused a backlash and led to vastly tightened restrictions. After 9/11, the "push" returned,

[12] An article entitled "Successful or Unsuccessful Operation" was included in *Echo of the Epics*, one of the group's online magazines, released February 14, 2010 (translation provided by subscription by the Search for International Terrorist Entities (SITE) Intelligence Group, http://news.siteintelgroup.com/).

with the restrictions on domestic security being viewed as one reason for the failure to prevent the attacks.

Although such swings might be seen as inevitable, they are unproductive and disruptive. Government agencies make investments, train personnel, and put programs in place to execute their missions within the constraints of law and policy. When those constraints shift, loosening after a major incident or tightening when public concern about abuses of power arise, organizations must restructure; individual careers can be damaged; money must be spent ending, changing, and reorganizing programs; and so on. Just as a short-term investor must pay to shake up a portfolio in response to near-term shifts in public opinion or market confidence, agency restructuring involves high transaction costs—costs that take away resources that could be devoted to improving security.

A deliberate rather than an urgency-driven approach to security planning is also particularly important for interventions that affect the civil liberties, privacy, and freedoms of individuals or that set up the potential for security organizations—even in well-meaning efforts to address real risks—to overstep lines and produce public or legal backlash. The urgency-driven approach risks adding momentum to the swinging pendulum, with potentially serious disruptions to our national homeland and counterterrorism policies.

America must build a homeland security strategy that is sustainable over the long term—one that politics and public reaction will not demand be inefficiently reshuffled and remade, even in the event of a future attack. Rather than viewing the pendulum swings between more security and less security as inescapable, policymakers should take advantage of periods of reduced urgency to design security policies that will help to curtail or reduce the likelihood of those swings, because the enduring goal—in fearless times as well as in fearful ones—is a domestic and international counterterrorism effort whose security returns will accrue over time, at reasonable and acceptable public and private costs, and whose initial investments can be built upon and improved to compound their returns even as the terrorist threats we face today increase, shift, decrease, or even are replaced with wholly different threats tomorrow.

## Related Reading

Brück, Tilman, "An Economic Analysis of Security Policies," *Defence and Peace Economics*, Vol. 16, No. 5, 2005, pp. 375–389.

Carafano, James Jay, *Homeland Security 3.0: Building a National Enterprise to Keep America Free, Safe and Prosperous*, Washington, D.C.: Heritage Foundation and the Center for Strategic and International Studies, 2008. As of June 17, 2011:
http://csis.org/publication/homeland-security-30

Committee on Science and Technology for Countering Terrorism, National Research Council, *Making the Nation Safer: The Role of Science and Technology in Countering Terrorism*, Washington, D.C.: National Academies Press, 2002. As of June 17, 2011:
http://www.nap.edu/catalog.php?record_id=10415

Committee on Technical and Privacy Dimensions of Information for Terrorism Prevention and Other National Goals, National Research Council, *Protecting Individual Privacy in the Struggle Against Terrorists: A Framework for Program Assessment*, Washington, D.C.: National Academies Press, 2008. As of June 17, 2011:
http://www.nap.edu/catalog.php?record_id=12452

Farrow, Scott, "The Economics of Homeland Security Expenditures: Foundational Expected Cost-Effectiveness Approaches," *Contemporary Economic Policy*, Vol. 25, No. 1, January 2007, pp. 14–26.

Flynn, Stephen E., "America the Resilient: Defying Terrorism and Mitigating Natural Disasters," *Foreign Affairs*, Vol. 87, No. 8, March/April, 2008, pp. 2–8.

Jackson, Brian A., *Marrying Prevention and Resiliency: Balancing Approaches to an Uncertain Terrorist Threat*, Santa Monica, Calif.: RAND Corporation, OP-236-RC, 2008. As of May 24, 2011:
http://www.rand.org/pubs/occasional_papers/OP236.html

Jackson, Brian A., Peter Chalk, Kim Cragin, Bruce Newsome, John V. Parachini, William Rosenau, Erin M. Simpson, Melanie W. Sisson, and Donald Temple, *Breaching the Fortress Wall: Understanding Terrorist Efforts to Overcome Defensive Technologies*, Santa Monica, Calif.: RAND Corporation, MG-481-DHS, 2007. As of May 24, 2011: http://www.rand.org/pubs/monographs/MG481.html

Jackson, Brian A., Agnes Gereben Schaefer, Darcy Noricks, Benjamin W. Goldsmith, Genevieve Lester, Jeremiah Goulka, Michael A. Wermuth, Martin C. Libicki, and David R. Howell, *The Challenge of Domestic Intelligence in a Free Society: A Multidisciplinary Look at the Creation of a U.S. Domestic Counterterrorism Intelligence Agency*, Santa Monica, Calif.: RAND Corporation, MG-804-DHS, 2009. As of May 24, 2011: http://www.rand.org/pubs/monographs/MG804.html

Klitgaard, Robert, Paul C. Light, eds., *High-Performance Government: Structure, Leadership, Incentives*, Santa Monica, Calif.: RAND Corporation, MG-256-PRGS, 2005. As of May 24, 2011: http://www.rand.org/pubs/monographs/MG256.html

Schneier, Bruce, *Beyond Fear: Thinking Sensibly About Security in an Uncertain World*, New York: Copernicus Books, 2003.

CHAPTER TWELVE

# Flight of Fancy? Air Passenger Security Since 9/11

*K. Jack Riley*[1]

*Jack Riley is vice president and director of the RAND National Security Research Division. He previously served as associate director of RAND Infrastructure, Safety, and Environment and director of the RAND Homeland Security Center.*

The phrase "touch my junk" became part of the lexicon of air passenger security in late 2010 thanks to the controversial decision by the U.S. Transportation Security Administration (TSA) to increase the physical scrutiny of air travelers. John Tyner, attempting to fly from San Diego, uttered the now-famous words when he refused to walk through a whole body image (WBI) scanner and subsequently also refused to submit to a full-body frisk. The latter would have involved a TSA agent touching his "junk," or genitals.

That national attention focused on the anatomy of a private citizen is one small indication of how distracted the country has become in its efforts to ensure air passenger security. The focus of these efforts should be on what works best and at least cost, but the traveling public remains doubtful that the focus has been put in the proper place. Although many travelers do not object to the use of WBI scans and

---

[1] I am indebted to the many individuals who contributed ideas to this essay, including Bob Poole of the Reason Foundation and the vocal and informed participants on Flyertalk's Travel Safety and Security forum. I am also grateful to the anonymous travelers who have given their opinions about airport security issues in lines and lounges and on planes all over the world.

frisks, many others do, grumbling about airport security issues in lines, lounges, and planes around the world.

TSA has moved toward this kind of screening in reaction to the December 2009 attempted bombing of an airplane over Detroit by Umar Farouk Abdulmutallab, the so-called "Christmas bomber," who had concealed explosive material in his underwear. The intent of TSA is to make WBI and frisks the primary methods by which all passengers are screened.[2] Those selected for a WBI scan who are unwilling or unable to complete it will have a TSA agent touching private parts of the body as part of intensified pat-downs.

The objections to the WBI scanning are varied.[3] The machines do not detect explosives but use x-rays or millimeter waves to generate an image of a person's body. Denser objects—such as metal, multiple folds of clothing, medical devices, or diapers—show up as darker areas (called anomalies) on the image. There is debate as to whether the machines, had they been in use, would have detected the explosive material that Abdulmutallab had concealed.[4] Health concerns have also been raised about these backscatter machines, which generate their images from x-rays.[5] Some travelers object to the slow speed of the WBI scans, which are demonstrably slower than earlier screening methods. Other objections include the fact that travelers with external medical

---

[2] It is not possible to reliably estimate the fraction of travelers on which these methods are being used. TSA has deployed approximately 400 WBI scanners, but they are not always in use at the airports where they are deployed.

[3] Pilots and flight crews objected when the machines were first introduced. Pilots have now been exempted.

[4] Steve Lord, *TSA Is Increasing Procurement and Deployment of the Advanced Imaging Technology, but Challenges to This Effort and Other Areas of Aviation Security Remain*, testimony before the Subcommittee on Transportation Security and Infrastructure Protection, Committee on Homeland Security, House of Representatives, Washington, D.C.: U.S. Government Accountability Office, GAO-10-484T, March 17, 2010, p. 9.

[5] In April 2010, four researchers from the Department of Biochemistry and Biophysics at the University of California, San Francisco, made their concerns public in a "letter of concern" to the Assistant to the President for Science and Technology (as of May 27, 2011: http://www.npr.org/assets/news/2010/05/17/concern.pdf). Health concerns are magnified by the lack of transparency in the maintenance schedule (Allison Young and Blake Morrison, "TSA to Retest Airport Body Scanners for Radiation," *USA Today*, March 14, 2011).

devices (including wheelchairs) or certain medical conditions (including a need for portable oxygen or an inability to stand without a cane or similar implement) are ineligible for such scans and instead must be patted down.[6] At many airports, WBI machines are configured in a way that creates personal insecurity: Travelers standing in the WBI cannot watch their personal goods go through the scanning machines, creating a risk of theft as goods exit the machines.[7]

Further objections arise from the fact that the "anomalies" detected are frequently false alarms that must be resolved with a pat-down.[8] Finally, many people object, usually on Fourth Amendment (unreasonable search and seizure) or privacy grounds, to the fact that the WBI scans generate detailed images of the traveler's body that are viewed—and perhaps stored—by government employees.[9] TSA has begun testing software modifications that use cartoon body images

---

[6] The TSA letter provides a list of all the conditions that prohibit use of the WBIs. It can be found, as of May 27, 2011, at: http://www.cfnews13.com/static/articles/images/documents/TSA-letter-to-disability-community-1123.pdf

[7] This is a problem with the backscatter machines in particular, as they have solid, non-transparent walls. The millimeter-wave machines have clear walls that better enable passengers to track personal items. However, some machines are configured in a way that prevents passengers from maintaining constant visual contact with their goods.

[8] Perhaps the highest-profile recent example comes from Sharon Cissna, a representative in the Alaska legislature. A previous bout with breast cancer left her with scarring that was visible on the WBI and therefore led to pat-downs. An account of her recent experience can be found at Scott McMullen, "TSA Bars AK State Rep. Sharon Cissna from Flying," Alaska TravelGram blog, February 21, 2011. As of May 27, 2011: http://www.alaskatravelgram.com/2011/02/21/tsa-bars-ak-state-rep-sharon-cissna-from-flying/)

TSA does not publish statistics on anomaly detection rates, but reporting on a test of machines deployed in Germany indicates high false-alarm rates from folded clothing, pleats, and other issues (Ulrich Gassdorf, *Hamburger Abendblatt*, February 11, 2011).

[9] It is unclear whether the machines can store images. Specifications in the procurement documents seem to require the machines to be able to store images, but TSA reports that they can store images only when the machine is in "test" mode (letter from Gale D. Rossides, TSA Acting Administrator, to Rep. Bennie G. Thompson, February 24, 2010; as of May 27, 2011: http://epic.org/privacy/airtravel/backscatter/TSA_Reply_House.pdf). In August 2010, the U.S. Marshals Service (part of the Department of Justice) acknowledged that more than 30,000 images were stored on a machine used at a courthouse in Orlando (letter from William E. Bordley, Associate General Counsel/FOI/PA Officer, Office of the General Counsel, U.S. Marshals Service, to John Verdi, Esq., Electronic Privacy Infor-

while still showing anomalies. While these methods address some privacy concerns, they do not alleviate the Fourth Amendment, health, effectiveness, speed, personal insecurity, or false-alarm concerns.

In addition to the growing discontent with this new form of security, there are looming financial constraints. Passenger security is one of the most visible and expensive components of the overall transportation security system. Given the costs associated with the new screening methods—an estimated $1.2 billion per year in capital, equipment, and operating costs by 2014[10]—it is worthwhile to ask three questions: How has passenger security performed since 9/11? What opportunities and innovations have been missed or not pursued? What steps should the United States be considering?

## Air Travel Is Safe and Secure

In the ten years since 9/11, about 6 billion enplanements (instances of passengers boarding a commercial plane) have occurred in the United States, and perhaps an additional 14 billion have occurred worldwide. Among those 20 billion passengers, according to the Aviation Safety Network, 7,019 died in aviation accidents between 2001 and 2009.[11] In addition, roughly 200 more have died on airplanes or at airports as a result of terrorism since 9/11.[12] Terrorists have succeeded in bringing

---

mation Center, August 2, 2010; as of May 27, 2011: http://epic.org/privacy/body_scanners/Disclosure_letter_Aug_2_2010.pdf). The Electronic Privacy Information Center filed a declaration against DHS in the U.S. District Court for the District of Columbia on May 27, 2010, that, among other requests, sought to compel the Department of Homeland Security (TSA) to produce 2,000 images stored for training purposes. This request was blocked on January 12, 2011, on the grounds that it would reveal vulnerabilities of the body-scanning technology.

[10] Mark G. Stewart and John Mueller, *Risk and Cost-Benefit Analysis of Advanced Imaging Technology Full Body Scanners for Airline Passenger Security Screening*, University of New Castle, Australia, Research Report No. 280.11.2010, January 2011.

[11] See the "Statistics" web page at the Aviation Safety Network website. As of May 27, 2011: http://www.aviation-safety.net/statistics

[12] National Consortium for the Study of Terrorism and Responses to Terrorism, Global Terrorism Database, no date. As of May 27, 2011: http://www.start.umd.edu/gtd/

down planes since 9/11, but they have been incapable of extending the damage by hijacking a plane and using it against a supplemental target. And since 9/11, no passengers have died in terrorist acts from enplanements originating in the United States. In short, despite the tragedy and loss of life on 9/11, air transportation is overwhelmingly a secure means of transportation, especially in the United States.

At least three factors contribute substantially to the improved security since 9/11.[13] First, and perhaps most importantly, passengers know now that they must be vigilant. The attacks of 9/11 were a wake-up call for a traveling public previously unaccustomed to suicide attacks. The lessons of 9/11 were learned within minutes of the first attack, as evidenced by the fact that the passengers and crew on Flight 93 realized that the attackers were on a suicide mission and fought to regain control of the plane.[14] Since then, passengers and crew have helped disrupt a number of additional incidents, including the attempted attacks by Richard Reid (the "shoe bomber") in 2001 and by Abdulmutallab.

Second, airlines have reinforced cockpit doors in a way that strictly limits access to the cockpit. Most crews have also modified their procedures to ensure that there is a barrier between the cockpit door and the passengers when the cockpit door needs to be opened. These steps mean that it is much more difficult, if not impossible, to commandeer an airplane and to conduct an attack similar to those of 9/11.

Third, changes to the visa approval process represent a relatively unheralded but important contribution to air transportation security. All 19 of the terrorists involved in the 9/11 attacks were in the United States on legitimate visas. Hence, the visa process was one target of

---

[13] Another augmentation of security has probably crept in, albeit inadvertently: An increasing fraction of domestic enplanements consists of small regional jets that are incapable of causing the kind of damage that occurred with the larger planes hijacked on 9/11.

[14] See, for example, Susan Sward, "The Voice of the Survivors: Flight 93, Fight to Hear Tape Transformed Her Life," *San Francisco Chronicle*, April 21, 2002. See also pp. 10–14 of the 9/11 Commission Report (*Final Report of the Commission on the Intelligence Capabilities of the United States Regarding Weapons of Mass Destruction*, Washington, D.C.: U.S. Government Printing Office, 2005. As of May 24, 2011: http://www.gpoaccess.gov/wmd/index.html

policy reforms.[15] Currently, residents of 36 countries can travel to the United States without obtaining a visa, but those traveling from other nations, such as Pakistan, Saudi Arabia, and Yemen, must obtain one.[16] Obtaining a visa involves providing extensive documentation (of the individual and, in some cases, family members, business associates, and the sponsor) that is investigated using homeland security, intelligence, and law enforcement databases and resources. Applicants also undergo an in-person consular interview. Overall, the number of nonimmigrant visas granted to residents of such countries fell sharply after 9/11 and, with some exceptions, has remained well below the number recorded in 2001.[17] In addition, the number of visa denials because of suspected links to terrorism increased from 47 (none of which were overturned as a result of subsequent evaluation) in 2002 to 683 (387 of which were overturned after additional investigation, leaving 296) in 2010. Thus, there has been a greater than sixfold increase in visa denials because of suspected links to terrorism.

Granted, the dramatic increase in visa denials entails costs beyond the visa process itself. By making it harder to come to the United States, we in America deter not only terrorists but also a large number of legitimate travelers—foreign tourists, foreign students, and qualified foreign workers—whose presence provides great benefits to our economy and the vibrancy of our culture. This opportunity cost needs to be accounted for when considering the security value of visa changes.

In the meantime, the three improvements cited above should help keep future aviation security incidents from being as catastrophic as the 9/11 attacks. In fact, future incidents are more likely to be on the scale of aviation *safety* incidents rather than *security* incidents. Thus, even if future attacks are successful, they need not lead to the loss of

---

[15] *Final Report of the Commission on the Intelligence Capabilities of the United States Regarding Weapons of Mass Destruction*, Washington, D.C.: U.S. Government Printing Office, 2005. As of May 24, 2011: http://www.gpoaccess.gov/wmd/index.html

[16] TravelState.gov, "Visa Waiver Program (VWP)," no date. As of May 27, 2011: http://travel.state.gov/visa/temp/without/without_1990.html#countries

[17] These statistics come from the FY2010 and FY2002 reports of the U.S. Department of State's Visa Office, Tables XVII and XX. As of May 27, 2011: http://travel.state.gov/visa/statistics/statistics_1476.html

confidence in the air travel system that the American public experienced on 9/11.

Despite the improvements made to air transportation security involving passengers, cockpits, and visas, the vast majority of U.S. transportation security money is spent at the airport to prevent passengers from bringing potentially dangerous goods on board. Although there are many prohibited items, including firearms and knives exceeding a certain length, the primary concern is with explosives. TSA annually spends about $5 billion on a workforce numbering an estimated 60,000. These screeners man the walk-through metal detectors, operate the baggage x-ray systems, search for liquids that exceed the established thresholds, man the WBI devices, conduct pat-downs, implement behavioral profiling, and conduct other screening functions. One reason for the workforce and expenses being so large is the fact that the screening functions are imposed virtually uniformly on every traveler entering an airport in the United States.

## Missed Opportunities

The high level of scrutiny to which U.S. airline passengers are subjected is a curious departure from the levels implemented in other areas of transportation and border security that use more risk-based approaches, such as the highly selective screening of shipping containers and the limited placement of federal air marshals aboard U.S. commercial flights. Until recently, though, the starkest contrast of all to the treatment of U.S. airline passengers was the treatment of U.S. commercial cargo on those *same flights* boarded by the passengers. It was not until August 2010 that all of the commercial cargo loaded onto domestic passenger planes was scanned or searched. Shipments on cargo jets from international destinations are not all currently screened, although the date for implementing the plan for screening them has been moved up from 2013 to 2011.

Another departure from the "inspect everyone and everything" approach involves TSA's own workforce at the airports. TSA employees are not screened when they enter the secure area of an airport throughout the course of the day, because they are trusted employees who have had a background check. They are thought to be at particularly low risk for coercion or conversion to radicalism. At many airports, certain other employees also have all-access badges that allow them to bypass security.

Thus, what is considered the "sterile area" of the airport is in fact not sterile. Substantial volumes of people (and, until recently, cargo) have made it into the sterile area without inspection. There have been no terrorist incidents associated with these leakages, however, suggesting that the *cargo* and *employee* risks have been appropriately managed for years (even before full cargo screening began) and begging the question of what kinds of risk management improvements might be available for *passengers*.

There are two main opportunities for improving the risk management system for passengers. First, flights originating in the United States are at much lower risk of being attacked by terrorists than are flights originating overseas. Second, enough is known about many passengers—their occupations, the security clearances they hold, their traveling profiles—to trust them to a greater degree than the current system does.

There is very little reason to be concerned about suicide bombers being present on flights *originating in the United States*. The security improvements noted above—passenger vigilance, cockpit security, and visa screening—go a long way toward preventing radical jihadists from entering the country or, having entered, from being able to commandeer a plane to conduct a spectacular attack. Moreover, the radical threat resident in and willing to conduct a suicide attack on the United States is extremely small.

Thus, the first opportunity for improving passenger risk management would be to differentiate between the domestic and international enplanements. One way to do this is to subject travelers who wish to come to the United States to a higher level of scrutiny than those already in the United States. This could be accomplished by maintaining current levels of inspection of travelers coming to the country and reducing the use of advanced equipment and intrusive methods inside the United States, where the threat is lower. Such a step would yield big savings in equipment and personnel by reducing the number of machines and agents required at U.S. airports. It would also reduce the deadweight losses that domestic travelers incur from arriving at airports early, waiting in lines, and undergoing intensive scrutiny.

The second opportunity, specifically for domestic enplanements (at least initially), would be to develop a trusted traveler program. The current security regime applies the same procedures to all 700 million passengers who board planes each year in the United States.[18] That we have not developed a reasonable way to reduce that inspection workload is perhaps the biggest missed opportunity of the past decade.

A trusted traveler program could be configured in a variety of ways. Recent conversations with airline industry executives suggest that a very small fraction of fliers account for a very large proportion of trips. In all likelihood, then, a trusted traveler program could be relatively small (with 5 million enrollees or less) and could still provide significant benefits. No program will be bulletproof, but such a program does not need to be given the extremely low odds of encountering a suicide terrorist on a flight originating in the United States. A trusted traveler program could initially be organized around these characteristics or combinations of characteristics:

- *Possession of a security clearance issued by a U.S. government agency.* Security clearances are issued after a comprehensive background investigation that includes an examination of foreign ties. These clearances are also far more stringent than the criminal background checks conducted on TSA agents. Individuals with security clearances are extremely unlikely to be involved in terrorist activities. The *Washington Post* reported in 2010 that more than 850,000 people held Top Secret clearances, which require an investigation covering the preceding ten years that includes contact with employers, co-workers, and others; involves investigation of education, employment, and personal and civic affiliations; and

[18] Bureau of Transportation Statistics, Research and Innovative Technology Administration, *Transportation Statistics Annual Report, 2010*, Table 2-2-5, "Domestic Enplanements at U.S. Airports: 1999–2009," p. 122.

includes agency checks of spouses and significant others.[19] Several additional million individuals hold Secret clearances that involve a similar level of investigation. The cost of these clearances has already been incurred, so the marginal cost of starting a trusted traveler program with this group would be low.

- *A profile that involves frequent travel.* An individual traveling 100,000 miles per year is, conservatively, spending 200 hours on airplanes a year. That is 10 percent of a standard 2,000-hour work year, suggesting that such travelers can be trusted with the basic screening that was in place prior to the deployment of WBI machines and pat-downs. Airlines generally do not make information on the size of their frequent flyer pools publicly available, but such individuals are thought to number in the tens to hundreds of thousands. Even at the lower end of the range, they would still be responsible for a large portion of the 630 million annual enplanements in the United States. The costs of a program based on frequent travel, and who would bear them, are unclear.
- *Willingness to submit to the equivalent of a security-clearance process.* Some travelers would find it well worth the time and expense to obtain such a credential in exchange for the ability to move through an airport more quickly. Several programs, including Global Entry, NEXUS, and SENTRI, already allow certain travelers to be pre-approved for expedited clearance for entry at U.S. borders. Global Entry members pay a fee, undergo an interview and background check, and provide fingerprints as part of seeking approval. SENTRI and NEXUS operate in a similar fashion at Mexican and Canadian ports of entry, respectively. The combined programs cover hundreds of thousands of frequent travelers. The marginal costs of implementing this approach would be relatively low, since more than a million travelers have already paid for these entry/exit credentials. Extending the privileges of these programs from entry into the United States to security at U.S. airports would be a relatively trivial and easily justified action.

---

[19] Dana Priest and William M. Arkin, "A Hidden World, Growing Beyond Control," in "Top Secret in America: A Washington Post Investigation," *Washington Post*, July 19, 2010.

I am not advocating that trusted travelers be *exempt* from security screening, in part because terrorists would make attempts to exploit the program. Rather, trusted travelers should be eligible for a level of primary screening that is not as restrictive, intrusive, and time-consuming as WBI and frisks—or even what was in place prior to the Abdulmutallab attempt. If U.S. trusted travelers were eligible for the pre-Abdulmutallab screening, they could see their processing further simplified by the development of special lanes where they would not be required to remove their shoes or computers and other electronics and where they would also be allowed to carry on liquids. These trusted traveler screenings would be supplemented by random applications of more intensive secondary screening to small portions of the trusted population. The random secondary screenings would help prevent contraband and risk from creeping into the process. In the meantime, the more intensive methods could be used more effectively on people about whom little is known.

Recognizing the security of flights originating in the United States and thus returning *all* passengers to the domestic procedures that existed before the recent additions would save, at minimum, about $1.2 billion annually. Additional savings could be achieved by eliminating the supplemental searches of passengers that now occur as they board planes and the use of roving teams to test passengers' beverages for explosive residue in the secure parts of airports.

The savings from a trusted traveler program would depend largely on how it was configured, on what fraction of the traveling public was qualified to be trusted travelers, and whether security procedures were already relaxed for U.S. enplanements. If procedures were relaxed for all U.S. enplanements, the incremental savings from a trusted traveler program would be smaller. Again, the savings would come from the need for marginally fewer personnel and machines. However, if procedures were not relaxed for all U.S. enplanements, the savings from a trusted traveler program could be substantially greater. The savings would rise with the fraction of travelers who are trusted and could easily approach those associated with relaxing the standards for all U.S. enplanements. Even greater savings could be achieved if trusted traveler status were something for which travelers had to pay.

Had these steps been implemented in the years after 9/11, the savings would now likely total in the tens of billions of dollars, with no discernible reduction in security.

## The Decade Ahead

Beyond their financial costs, the current screening methods, which are slower than those they replaced, impose additional losses on travelers, who could use the time they spend waiting for airport security more productively. In addition, the new security measures seem likely to deter some people from traveling at all and to push some toward using other modes of transportation. Deterring travel will impose additional losses on the economy, and travelers who are now choosing to drive instead of fly may be placing themselves at greater risk.

Researchers have estimated that the 9/11 attacks generated nearly 2,200 additional road traffic deaths in the United States through mid-2003 from a relative increase in driving and reduction in flying resulting from fear of additional terrorist attacks and associated reductions in the convenience of flying.[20] If the new security measures are generating similar, or even smaller, substitutions and the driving risk has grown as hypothesized, the new methods could be contributing to more deaths *annually* on U.S. roads than have been experienced *cumulatively* since 9/11 from terrorism against air transportation targets around the world.

Returning to the domestic air security procedures that had existed prior to December 2009 and creating a trusted traveler program are two relatively short-term steps that can be taken. What other changes should we be looking at for the longer term?

While by no means a trivial change, TSA should be required to analyze proposed security measures and regulations, using clear, transparent, peer-reviewed risk management principles. Congress can help in this regard by requiring TSA to use such methods in reporting on significant policy changes to passenger security. One reason for the cur-

---

[20] Garrick Blalock, Vrinda Kadiyali, and Daniel H. Simon, "Driving Fatalities After 9/11: A Hidden Cost of Terrorism," *Applied Economics*, Vol. 41, No. 14, 2009, pp. 1717–1729.

rent situation is that security measures have been grafted on or layered on in response to specific incidents, with little regard to an integrated assessment of cost, effectiveness, and impact. Risk management modeling can be used to assess these variables.[21] It is possible to calculate how much a policy would need to reduce the annual losses from terrorism to cover the cost of implementing it. For example, a recent study, using conservative assumptions, found that WBI machines would have to disrupt more than one attack involving body-borne explosives *and* originating from U.S. airports every two years to be cost-effective.[22] Given that these joint conditions are not currently present, it is fair to conclude that the new methods are not worth the tradeoff of public expenditures and costs to travelers.

As noted in other chapters in this volume, terrorists are unlikely to go away, and they seem intent on developing new and more inventive ways of disrupting our society. Their intent, however, does not justify the blind application of restrictive security measures that impede commerce, compromise privacy, and imperil civil liberties. Yet, for most of the past decade, the United States has pursued policies with very little regard to the costs they impose on travelers or the net reduction in risk that they generate. It is imperative that we use the next decade to develop smarter, more sustainable, and more practical solutions to air passenger security. We should start by rolling back the procedures that were implemented in late 2010, which appear to exceed any reasonable test of regulatory cost-effectiveness. Further savings can likely be gained by subjecting other security measures—such as shoe removal, the ban on liquids, gate inspections, and the use of behavioral-detection officers—to careful scrutiny.

---

[21] *Review of the Department of Homeland Security's Approach to Risk Analysis*, Committee to Review the Department of Homeland Security's Approach to Risk Analysis, National Research Council, 2010. As of June 21, 2011: http://www.nap.edu/catalog.php?record_id=12972

[22] Mark G. Stewart and John Mueller, "Risk and Cost-Benefit Analysis of Advanced Imaging Technology Full Body Scanners for Airline Passenger Security Screening," University of New Castle, Australia, Research Report No. 280.11.2010, January 2011.

## Related Reading

Riley, K. Jack, Bruce W. Bennett, Mark Hanson, Stephen J. Carroll, Lloyd Dixon, Scott Gerwehr, Russell W. Glenn, Jamison Jo Medby, and John V. Parachini, *The Implications of the September 11 Terrorist Attacks for California: A Collection of Issue Papers*, Santa Monica, Calif.: RAND Corporation, IP-223-SCA, 2002. As of May 24, 2011: http://www.rand.org/pubs/issue_papers/IP223.html

Stevens, Donald, Thomas Hamilton, Marvin Schaffer, Diana Dunham-Scott, Jamison Jo Medby, Edward W. Chan, John Gibson, Mel Eisman, Richard Mesic, Charles T. Kelley Jr., Julie Kim, Tom LaTourrette, and K. Jack Riley, *Implementing Security Improvement Options at Los Angeles International Airport*, Santa Monica, Calif.: RAND Corporation, DB-499-1-LAWA, 2006. As of May 24, 2011: http://www.rand.org/pubs/documented_briefings/DB499-1.html

U.S. Government Accountability Office, *TSA Is Increasing Procurement and Deployment of the Advanced Imaging Technology, but Challenges to This Effort and Other Areas of Aviation Security Remain*, Washington, D.C.: U.S. Government Printing Office, March 17, 2010. As of May 24, 2011: http://www.gao.gov/new.items/d10484t.pdf

Wilson, Jeremy M., Brian A. Jackson, Mel Eisman, Paul Steinberg, and K. Jack Riley, *Securing America's Passenger-Rail Systems*, Santa Monica, Calif.: RAND Corporation, MG-705-NIJ, 2007. As of May 24, 2011: http://www.rand.org/pubs/monographs/MG705.html

CHAPTER THIRTEEN

# The Intelligence of Counterterrorism

*Gregory F. Treverton*

*Gregory Treverton, a former vice chair of the National Intelligence Council, directs the RAND Corporation's Center for Global Risk and Security. A second edition of his* Intelligence for an Age of Terrorism *will be published later this year by Cambridge University Press.*

September 11, 2001, marked a sea change for U.S. intelligence, one that is widely acknowledged but far from fully grasped. The reversal in the priority of targets—from nation-states to transnational groups, such as terrorists—is widely acknowledged. But this change goes to the heart of how intelligence does its business: from collection to analysis to dissemination. The change is nicely captured in a line given to me by a young analyst at the U.S. National Geospatial Intelligence Agency: "Imagery used to know what it was looking for and be looking for things; now it doesn't know what it's looking for and is looking not for things but rather for activities."

In the case of Osama bin Laden, of course, the United States *did* know what it was looking for. Nonetheless, the quest for bin Laden also illustrates how much the target has changed since 9/11, and his killing underscores what has become the intensely operational nature of the counterterrorism intelligence task.

The shift in priority from nations to groups or even individuals as primary targets would indicate that the overall threat to U.S. national security has diminished. Yet the perceptions have been far different. The Soviet Union was much more dangerous than any terrorist group, but at least the Soviet Union was an enemy we could know. It had borders and could be assumed to behave rationally, in our terms. The relative obscurity of the terrorist enemy, in contrast, might help explain why Americans fear it well beyond the real danger so far. Moreover, politicians inadvertently abet those fears by seeming to imply that *any* terrorist attack can be prevented. Terrorists' work is sowing terror, and we in America often do it for them.

Intelligence naturally focuses on new threats and worst cases and so contributes to the fear without really meaning to do so. It is time for U.S. intelligence agencies to rethink not only how they do their business but also how they can do that business while adding less to the climate of terror that gives the terrorists a win.

## From Nation-States to Terrorist Targets

Nation-states are geographic; they have addresses. As important, they come with lengthy "stories" attached, and intelligence is ultimately about helping people adjust the stories in their heads to guide their actions. Absent some story, new information is just a factoid. We know what states are like, even states as different from the United States as North Korea. They are hierarchical and bureaucratic. They are a bounded threat. Many of the items of interest about states are big and concrete: tanks, missiles, massed armies.

Terrorists are different in every respect. They are small targets, like bin Laden, yet a single suicide bomber can do major mayhem. They are amorphous, fluid, and hidden, presenting intelligence with major challenges simply in describing their structures and boundaries. Not only do terrorists not have addresses, they aren't only "over there." They are "here" as well, an unpleasant fact that impels nations to collect more information on their citizens and residents and to try to do so with minimal damage to civil liberties. Terrorists come with little story attached. A decade after 9/11, we still debate whether al Qaeda is a hierarchy, a network, a terrorism venture capitalist, or an ideological inspiration. No doubt it contains elements of all four, but that hardly amounts to a story.

Ironically, the secretive foes of the Warsaw Pact were easier to figure out from afar than are the terrorists in our midst. Cold War intelligence gave pride of place to secrets—those precious nuggets of information, gathered by human and technical means, that intelligence "owned." In contrast, an avalanche of data is available on terrorists; witness the 9/11 hijackers whose true addresses were available in California motor-vehicle records. But the sheer volume of data, plus the lack of a story with parameters, means that information-gathering against terrorists necessarily involves "mining" or other processing of large quantities of information. The hardest terrorists of all to pin down are the near lone wolves, like U.S. Army Major Nidal Hassan, the Fort Hood killer of 2009.

Another difference is that terrorists constantly adapt to their targets. Former U.S. Secretary of Defense Harold Brown once quipped about the U.S.-Soviet nuclear competition, "When we build, they build. When we stop, they build." While the United States hoped to influence Moscow, intelligence could presume that it would not—the Soviet Union would do what it would do. The terrorist target, however, is utterly different in this respect. It is the ultimate asymmetric threat, shaping its capabilities to our vulnerabilities.

The 9/11 suicide bombers did not decide on their attack plan because they were airline buffs. They had done enough tactical reconnaissance to know that fuel-filled jets in flight were vulnerable assets and that defensive passenger-clearance procedures were weak. Thus, to a great extent, we shape the threat to us; it reflects our vulnerable assets and weak defenses. This interaction between "us" and "them" has very awkward implications for U.S. intelligence, especially for agencies such as the CIA, which have "foreign" missions and thus have traditionally been enjoined from doing domestic intelligence.

A final major difference between transnational targets, especially terrorists, and state targets may be the most important of all. If *preventing a terrorist attack* is the name of the game, the pressure on intelligence is extraordinary. The dominant strategy of the Cold War—deterrence—was less sensitive to the specifics of intelligence. It rested on the assumption that, for all its differences in goals and ideology, the Soviet Union was like us: modern, rational (in our terms), and not self-destructive. In contrast to deterrence, prevention of suicidal attacks—whether by preemption, disruption, or simply defending vulnerabilities—requires enormous precision in intelligence, even to the point of understanding individual intentions.

## How Are We Doing?

In retrospect, we exaggerated the terrorist threat after 9/11. For those of us who had worked on terrorism prior to 9/11, the attack came as a shock but not a surprise. The shock was not the attack itself—a series of blue-ribbon panels had warned that terrorism inevitably would come to our shores—but the terrorists' ability to mount four coordinated strikes.

In the five years after 9/11, fewer than 100 American civilians were killed each year by terrorists worldwide. During the same five years, an average of 62 people were killed each year by lightning, 63 by tornadoes, 692 in bicycle accidents, and 41,616 in motor-vehicle–related accidents. That terrorism frightens Americans well beyond the reality owes not just to analysts focusing on worst cases but also to politicians hyping the threat.

When President Obama was caught telling an interviewer that America could "absorb another terrorist attack," there were some who immediately accused him of callousness or worse.[1] However, the criticism was surprisingly limited and short-lived. This could be an important indication that the public is more ready to hear this message than some of our leaders imagine.

Hyped threat and finger-pointing aside, the nation's intelligence has made progress since 9/11. Those attacks were blamed on a failure to "connect the dots." Foiling that plot would have required not just creative leaps of foresight by intelligence analysts but also political will to take draconian measures to prevent a large-scale attack on U.S. soil,

---

[1] "United States Could 'Absorb' Another Terror Attack, Obama Says in Woodward Book," *Fox News*, September 22, 2010. As of June 17, 2011: http://www.foxnews.com/politics/2010/09/22/obama-divided-afghan-war-woodward-book/

something that hadn't happened since Pearl Harbor and was therefore almost unthinkable. By 2009, in contrast, Umar Farouk Abdulmutallab (the "underwear bomber") and his Yemeni helpers were on the U.S. radar screen. Simply singling him out for a body search would have done the job. The intelligence community did not connect the dots, so the fact that his attack failed was due to luck, but at least this time there were dots.

So far, the three major initiatives since 9/11 to improve intelligence against the terrorist threat might be characterized as "worse than expected," "better than expected," and "as hard as expected."

The first, "worse than expected," is the 2004 creation of the position of director of national intelligence (DNI). The blue-ribbon panel that investigated 9/11 argued artfully that the absence of a person with overall responsibility for coordinating the nation's intelligence capabilities contributed to the failures that led to 9/11. In fact, the failures owed more to the absence of coordination at the *working* level than to the absence of broad strategic direction, and the bare fact that 9/11 occurred was enough to spur marked improvement in day-to-day coordination. This collaboration proved to be particularly impressive in the killing of bin Laden, which demonstrated coordinated "geocells" in action—that is, the overlaying of signals intelligence and imagery with information from informants to find the trail of bin Laden and then make his hideout transparent to U.S. raiders.

Still, the need for better strategic management of the far-flung intelligence community did argue for a DNI. However, eleventh-hour compromises with the defense committees of the U.S. Congress cut back the DNI's authority, leaving the first directors to concentrate not on managing the 17 intelligence agencies but on being the president's senior intelligence adviser. DNIs have since received a lot of what Washington calls "push back" from other agencies on issues large and small. As a result, it is difficult to gainsay those critics who feared the DNI would be just another layer of bureaucracy. It is cautionary that when the third DNI, Admiral Dennis Blair, feuded first with President Obama's CIA director, Leon Panetta, over who would control what had been CIA stations abroad and then with the White House staff over closer intelligence relations with France, it was Blair who lost and departed in 2010.

In the "better than expected" category is the most important but least noticed initiative of the three: the reordering of priorities at the FBI from law enforcement to intelligence-led prevention. The 9/11 attacks prompted immediate calls for the establishment of a new domestic intelligence service, separate from the FBI. The 9/11 Commission's diagnosis pointed straight at the limitations imposed by the FBI's culture of case-based law enforcement, saying that FBI agents were "trained to build cases, [and] developed information in support of their own cases, not as part of a broader more strategic [intelligence] effort."

But FBI Director Robert S. Mueller, who had been in the post for just one week on 9/11, moved quickly to reorient the bureau from building cases toward gathering intelligence, sending a reorganization plan to Congress in November 2001. The bureau's top priority became, in the words of its website, to "protect the United States from terrorist attack." In an organization where intelligence analysts were once lumped with all other non-agents as "support" (sometimes dubbed "furniture"), there are now field intelligence groups in each field office analyzing and disseminating intelligence—a dramatic change if still very much a work in progress.

In the "as hard as expected" category is what is called "information-sharing." At the federal level, the DNI's National Counterterrorism Center (NCTC) is responsible for "connecting the dots" across intelligence agencies. It represents an innovation from the Cold War approach to structuring intelligence, in which collection was organized around *sources*—signals, espionage, and imagery—but analysis was organized by each individual *agency*. That approach made some sense when there were few collectors and one main secretive target. In effect, the collectors were asked, What can you contribute to understanding the puzzle of the Soviet Union? That approach makes no sense now, with countless sources of information about obscure targets, and NCTC is a worthy effort to structure intelligence around an issue or problem, not a source or agency. NCTC deserves high marks for serving senior policymakers.

But the intelligence effort against terrorism today needs to include the eyes and ears of some 700,000 law enforcement officers in 18,000 law enforcement agencies, as well as private sector managers of critical infrastructure. The number of FBI-sponsored joint terrorism task forces (JTTFs), which predated 9/11, has expanded to more than 100 around the country. Meanwhile, the U.S. Department of Homeland Security is promoting a newer initiative, called "fusion centers," to assemble strategic intelligence at the regional level.

Like JTTFs, the fusion centers bring together federal, state, and local officials and involve the private sector. They are very much a work in progress. Communication among them ranges from poor to absent, and not all have statewide intelligence systems. They also do not all have access to law enforcement data or private sector information. The lack of interoperability among different agencies' information systems, widely criticized directly after 9/11, still exists. Most fusion centers are becoming all-crimes operations, especially in places where terrorism is a minor threat. Some will surely disappear as federal support winds down.

The intelligence challenges posed by terrorism remain daunting, but that does not mean the threat is overwhelming. Politicians and intelligence officials alike need to do a better job of speaking plainly to the American people. Terrorism, as long as terrorists do not obtain nuclear or highly destructive biological weapons, is more a nuisance than a mortal threat. America's security and intelligence apparatus can always do better, and it should. But it will never be able to stop every terrorist plot—a grim reality that Americans need to come to better grips with. Calling another attack "intolerable" is wishing, not making policy. Some honest talk would be useful, so that when the next major attack comes—as it surely will—it is not viewed as the end of the world.

## Related Reading

Davis, Lynn E., Gregory F. Treverton, Daniel Byman, Sara A. Daly, and William Rosenau, *Coordinating the War on Terrorism*, Santa Monica, Calif.: RAND Corporation, OP-110-RC, 2004. As of May 24, 2011: http://www.rand.org/pubs/occasional_papers/OP110.html

*Final Report of the Commission on the Intelligence Capabilities of the United States Regarding Weapons of Mass Destruction*, Washington, D.C.: U.S. Government Printing Office, 2005. As of May 24, 2011: http://www.gpoaccess.gov/wmd/index.html

National Commission on Terrorist Attacks Upon the United States, *The 9/11 Commission Report*, Washington, D.C.: U.S. Government Printing Office, 2004. As of May 24, 2011: http://www.gpoaccess.gov/911/index.html or http://www.9-11commission.gov/

Riley, K. Jack, Gregory F. Treverton, Jeremy M. Wilson, and Lois M. Davis, *State and Local Intelligence in the War on Terrorism*, Santa Monica, Calif.: RAND Corporation, MG-394-RC, 2005. As of May 24, 2011: http://www.rand.org/pubs/monographs/MG394.html

Treverton, Gregory F., *Intelligence for an Age of Terror*, Cambridge, UK: Cambridge University Press, 2009.

———, *The Next Steps in Reshaping Intelligence*, Santa Monica, Calif.: RAND Corporation, OP-152-RC, 2005. As of May 24, 2011: http://www.rand.org/pubs/occasional_papers/OP152.html

———, *Reorganizing U.S. Domestic Intelligence: Assessing the Options*, Santa Monica, Calif.: RAND Corporation, MG-767-DHS, 2008. As of May 24, 2011: http://www.rand.org/pubs/monographs/MG767.html

———, *Reshaping National Intelligence for an Age of Information*, Cambridge, UK: Cambridge University Press, 2001.

Treverton, Gregory F., and C. Bryan Gabbard, *Assessing the Tradecraft of Intelligence Analysis*, Santa Monica, Calif.: RAND Corporation, TR-293, 2008. As of May 24, 2011: http://www.rand.org/pubs/technical_reports/TR293.html

# PART FIVE
# Inspired to Build a Stronger America

There is a tendency within American history to draw inspiration from devastation, hope from misfortune, optimism from hardship. It is a tendency rooted in the American experience of having carved meaning and virtue out of a vast new frontier. It remains integral to the identity of this nation, founded and often reinvigorated by immigrants, to deem the struggle of displacement worthwhile for the sake of a better life for those who will follow. It is an abiding faith of a people who believe they are destined to shine a light, a beacon, a torch of freedom for the world.

The 9/11 attacks and their ostentatious murders of nearly 3,000 Americans devastated the country's spirit as much as its centers of financial and military power. The notion that there could be a silver lining to such hideous destruction might offend those who have suffered the most. But it would also be the ultimate triumph of terrorism to extinguish this flame of hope forever.

The three essays in Part Five speak of resilience. Of building a stronger America. Of using and overcoming the terrorist threat to make the country healthier, fairer, and truer to its ideals. Of taking the devastation, misfortune, and hardship and deriving inspiration, hope, and optimism.

Jeanne Ringel and Jeffrey Wasserman believe the 9/11 attacks, along with the anthrax attacks that soon followed, were "an important wake-up call for America's leaders, including the leadership of America's public health system." Signs of a turnaround emerged in 2002, "when senior federal officials and the U.S. Congress realized that a robust public health system is vital for national security." The recession has stalled progress, and numerous challenges remain. But Ringel and Wasserman point to examples, notably America's response to the H1N1 epidemic, to demonstrate how the tragedy of 9/11 has already helped to invigorate America.

The untamed legal landscape for compensating terrorist victims offers another patch of ground on which to fortify America, according to Lloyd Dixon, Fred Kipperman, and Robert Reville. "The compensation system for losses following another large attack remains undeveloped and highly uncertain. We are left with a system that provides clear signals neither on what actions private firms and public agencies should take to mitigate the risk of terrorism . . . nor on what losses they may be required to cover should another attack occur."

That means a compensation system can now be crafted to promote social cohesion and national unity, rather than legal and financial wrangling, in the event of future terrorist attacks. In this way, compensation policy can contribute to social and economic resiliency. Promoting solidarity through compensation policy could even help deter future terrorist attacks "by causing terrorism to be less effective in achieving its strategic goals of inciting fear and division."

America's far-reaching response to 9/11 "invited an assertion of executive authority, made security a national preoccupation, and tested the nation's commitment to American values," writes Brian Michael Jenkins. American values were damaged, he insists. "The most egregious example was the employment of coercive interrogation techniques that were tantamount to torture. The sophistic legal defense of these techniques remains a blight on American history."

At times, American values were abandoned, "but resilient congressional and judicial institutions, cognizant and at times jealously protective of their roles and prerogatives, have struggled mightily to right the ship of state and to chart a new course for the nation." It is what Jenkins calls "a messy but overall optimistic assessment." It is a quintessentially American way of moving forward.

CHAPTER FOURTEEN

# The Public Health System in the Wake of 9/11: Progress Made and Challenges Remaining

*Jeanne S. Ringel and Jeffrey Wasserman*

*Jeanne Ringel, a RAND senior economist, is leader of the Public Health Systems and Preparedness initiative within RAND Health and a faculty member of the Pardee RAND Graduate School. Jeffrey Wasserman, a RAND senior policy researcher, is assistant dean for academic affairs at the Pardee RAND Graduate School.*

The 9/11 attacks, along with the anthrax attacks that soon followed, were an important wake-up call for America's leaders, including the leadership of America's public health system. That system had become in part a victim of its own success by all but eradicating a string of devastating diseases, including smallpox and polio. Public health agencies, especially those at the state and local levels, were thought to be overstaffed and therefore were targeted for budget cuts. Things got so bad that a 1988 Institute of Medicine report bluntly concluded, "The public health system is in disarray." Fourteen years later, the situation had not improved. A follow-up 2002 report from the institute noted the lack of progress since 1988 and concluded, "The public health system remains in disarray today."

Signs of a turnaround, however, began to emerge in 2002, when senior federal officials and the U.S. Congress realized that a robust public health system is vital for national security. In response, the federal government, through the Centers for Disease Control and Prevention (CDC) and Health Resources and Services Administration,[1] signed a set of cooperative agreements with states, territories, and a handful of large cities to strengthen public health infrastructure and preparedness. These agreements injected about a billion dollars per year into state and local public health systems and hospitals.[2] This served as a potent vehicle for rebuilding what many critics believed had become a moribund public health system.

Initially, attention and funding were focused heavily on bioterrorism. The National Institutes of Health received an infusion of funds for basic research related to biodefense. The Public Health Security and Bioterrorism Preparedness and Response Act of 2002 mandated the development of BioSense, an integrated public health surveillance system for the early detection and rapid assessment of potential bioterrorism-related illnesses. Similarly, the U.S. Department of Homeland Security deployed the BioWatch Program, a network of sensors within existing Environmental Protection Agency air filters, to detect potential bioterror attacks. At the state and local levels, governments were required to develop plans for vaccinating all first responders against smallpox and for dispensing countermeasures to respond to an aerosolized anthrax attack.

Over time, however, it became apparent that it was necessary to consider a broader range of threats (including natural disasters, radiological incidents, and bombings), and there was a movement toward an "all hazards" approach to planning. Given the large array of threats to prepare for, an all-hazards approach that focuses on a core set of capabilities that are relevant across a wide range of threats is thought

[1] Initially, those cooperative agreements focusing exclusively on hospital preparedness were administered by the Health Resources and Services Administration, but they were subsequently transferred to the Office of the Assistant Secretary for Preparedness and Response.

[2] For detailed information on the amount of funding for the cooperative agreements, see Centers for Disease Control and Prevention, *Public Health Preparedness: Strengthening the Nation's Emergency Response State by State*, Appendixes 3 and 4. As of June 18, 2011: http://www.bt.cdc.gov/publications/2010phprep/; and Department of Health and Human Services, "HHS Fact Sheet: FY10 Hospital Preparedness Program (HPP)," no date. As of June17, 2011: http://www.phe.gov/preparedness/planning/hpp/pages/fy10hpp.aspx

by many to make the most effective use of scarce planning resources (both money and time).

With the infusion of funds into public health preparedness, public health officials, particularly at the state and local levels, started to broaden their roles from the traditional practice of public health. It became apparent that their expertise was needed to swiftly detect, contain, and minimize morbidity and mortality associated with a wide range of threats. This expanded mission put public health officials "at the table," literally and figuratively, along with such traditional first responders as police chiefs, fire captains, and emergency medical service directors. Since this turnaround, public health has consistently demonstrated its value in responding to health emergencies triggered by hurricanes, floods, infectious disease outbreaks, and oil spills.

Although progress has been made in many areas of public health policy since 9/11, important challenges remain. The various components of what can only loosely be called a public health *system* are still poorly integrated. Issues over who has authority to do what in a public health emergency remain unresolved, and considerably more attention must be paid to the ways in which resources are deployed.

Equally important, the large infusions of federal funding that followed 9/11 have decreased over time. Although the cooperative agreements still exist, the level of funding disbursed in 2010 was only three-quarters of that in 2003. The recession, which began in 2007, has dramatically compounded the problem by forcing many states and local governments to slash funding for government services, including public health.[3] The results have been devastating for many agencies.

To meet these challenges and continue to make progress, a significant cultural shift will be needed. The cultural shift will require that all sectors of society, not just the public health and emergency manage-

---

[3] National Association of County and City Health Officials, "Local Health Department Job Losses and Program Cuts, Findings from January/February 2010 Survey," Washington, D.C.: Research Brief, May 2010. As of June 17, 2011: http://www.naccho.org/topics/infrastructure/lhdbudget/loader.cfm?csModule=security/getfile&PageID=140774; Association of State and Territorial Health Officials, "Budget Cuts Continue to Affect the Health of Americans: Update May 2011," Research Brief, May 2011. As of June 17, 2011: http://www.astho.org/Display/AssetDisplay.aspx?id=6024

ment sectors, recognize their important roles in contributing to public health emergency preparedness, response, and recovery.

After describing some of the key successes that have been achieved and the challenges that remain, we discuss the necessary next steps. We believe that if this path is pursued in earnest, the tragedy of 9/11 can ultimately contribute to a stronger and more resilient America.

## Progress Has Been Made on Many Fronts

Over the past decade, federal, state, and local governments have made a wide array of changes with respect to how public health systems are organized and managed. There is now a much greater emphasis on "preparedness," which has come to be used as a shorthand term for activities that center on health threats posed by large-scale emergencies. Because of their size and scope, such emergencies hold the potential to overwhelm a community's ability to contain the damage and protect human health.

In the years after 9/11, major investments in preparedness were made by expanding and improving key response systems, such as disease surveillance, public health laboratories, and communications. Emergency-response training programs were developed to prepare a more flexible and capable public health workforce. Extensive planning was undertaken at the federal, state, and local levels to promote public health disaster response and recovery. As a result, detailed plans are now in place for mass delivery of medications and other medical supplies to communities when needed.

Similarly, starting in 2006, a substantial investment in pandemic preparedness was made, spurred in large part by concern about the H5N1, or avian flu, virus circulating in Asia.[4] With this investment, detailed pandemic influenza plans were put in place well ahead of the

[4] For a detailed discussion of pandemic preparedness planning during this time frame, see Stewart Simonson, "Reflections on Preparedness: Pandemic Planning in the Bush Administration," *Saint Louis University Journal of Health Law and Policy*, Vol. 4, No. 5, no date. As of June 17, 2011: http://law.slu.edu/healthlaw/journal/archives/Simonson_Article.pdf.

outbreak of H1N1 in April 2009. After Hurricane Katrina revealed inadequacies in preparedness planning, particularly for the evacuation and care of vulnerable populations, government agencies at all levels took additional steps to ensure that the needs of these populations are adequately addressed in preparedness and response plans.[5]

Substantial progress has also been made in developing valid and reliable performance measures[6] and in implementing continual quality improvement methods in health departments. Although it is difficult to say with absolute confidence that the United States is better prepared to confront large-scale health threats today than it was nine years ago, before the implementation of the cooperative agreements, evidence gathered from case studies, tabletop exercises, drills, and other data sources offers promise.

Perhaps the most substantial evidence of progress in building public health preparedness and response capabilities can be found in America's response to the recent H1N1 pandemic. Although the disease turned out to be less virulent than first feared, it nevertheless presented a formidable and sustained challenge to public health systems in the United States and around the world and offered a unique opportunity to assess the nation's response capabilities. The investments in planning and response capacity paid off in a number of areas: the quick identification and characterization of a novel pandemic virus, the record-breaking pace with which a new flu vaccine was developed and produced, the rapid distribution of antiviral drugs from the CDC's Strategic National Stockpile, and highly effective communications with the general public regarding methods to prevent transmission of the virus and to deal with its consequences.

---

[5] In fact, the Pandemic and All Hazards Preparedness Act (PAHPA) (PL 109-417) required that the U.S. Department of Health and Human Services integrate the needs of at-risk individuals into emergency planning.

[6] For example, the CDC Division of State and Local Readiness has developed a set of capability measures that are reported by states as a requirement of public health emergency preparedness cooperative agreements (Centers for Disease Control and Prevention, "Public Health Preparedness Capabilities: National Standards for State and Local Planning," no date. As of June 17, 2011: http://www.cdc.gov/phpr/capabilities/index.htm

## Important Challenges Remain

Despite progress on a number of important fronts, notable challenges remain.

While generally deemed successful, the response to H1N1 highlighted a number of areas where improvements are needed. Although the pandemic vaccine was produced in record time, it was still not widely available in time to have a significant impact on the epidemic. Despite the U.S. Department of Health and Human Services having spent more than $1 billion to enhance vaccine manufacturing capacity, newer technologies were not ready in time to play a role in the epidemic. The U.S. response, therefore, relied on production technologies that have not fundamentally changed in several decades, using egg-based production lines rather than the cell-based production lines used in Europe and elsewhere. New production technologies are needed to greatly reduce the time required to mass-produce a pandemic vaccine. The response to H1N1 also illustrated the need for more nimble methods of obtaining and distributing funds to support the response at the state, territorial, tribal, and local levels. While the federal government was able to mobilize substantial new funding to respond to the pandemic, the process for obtaining funds was lengthy and burdensome for state and local responders.

Hurricane Katrina exposed serious deficiencies in crisis decision-making, particularly the clear delineation of roles and responsibilities among government agencies at all levels, and in the coordination of government actions with nongovernmental organizations, private sector firms, and other stakeholders. The tragic loss of life sparked important advances in command and control, notably the adoption of the National Response Framework, although problems remain with coordination across federal departments and agencies. More recently, the Deepwater Horizon oil spill, which affected many communities that were still recovering from Hurricanes Katrina and Rita, demonstrated the need to involve the public health system early in any large-scale event that has the potential to affect human health.

In fact, advocates for what is known as "all hazards" preparedness can rightfully note that although the bulk of federal preparedness spending in the immediate aftermath of 9/11 and the anthrax attacks was directed at countering the threat of bioterrorism, the three biggest domestic challenges to public health since then have been a hurricane, a flu pandemic, and an oil spill. The massive loss of life that followed the earthquake in Haiti and the earthquake-spawned tsunamis in the Indian Ocean and Northeastern Japan (the latter tragically compounded by a reactor leak) illustrate the breadth and magnitude of threats faced by public health in the United States and abroad.

One of the most notable unmet challenges is the need to improve the ability of the health-care system to absorb a large-scale "surge" of ill or injured victims caused by a health emergency. Emergency rooms in many urban areas are regularly gridlocked with patients, so much so that the diversion of inbound ambulances—once unthinkable—is commonplace today. Many intensive-care units and hospital wards are filled to capacity. Hospital officials state that they could quickly open up beds by discharging a certain number of inpatients, but it is difficult to determine where those individuals would go.

Hospitals and other health-care facilities need to work with their local and state health departments and other response partners in their communities to develop and exercise plans to ensure that people receive the right levels of care in the right settings at the right time during and after a public health emergency. By working together, health-care facilities can leverage resources and provide capacity well in excess of what could be achieved with each facility working independently.

In the years since 9/11, the primary focus of preparedness efforts and investments has been on improving the capabilities to *respond* to a disaster; very little attention has been devoted to identifying and building the capabilities to enable communities to swiftly *recover* from one. Moreover, when attention has been paid to recovery, it has focused primarily on restoring physical infrastructure rather than assuring human recovery.[7] To bring a community back to a status that is roughly as good as, if not better than, its status before an incident, it is necessary

[7] Anita Chandra and Joie D. Acosta, "Disaster Recovery Also Involves Human Recovery," *Journal of the American Medical Association*, 2010, Vol. 304, No. 14, pp. 1608–1609.

to develop recovery plans that foster collaboration, build connections among neighbors, and meet the physical and behavioral health needs of the population, including disaster responders.

In the course of our work, we have observed troubling inequities in the degree to which communities are ready to cope with a myriad of health-related events. Over much of the past decade, through tabletop exercises, drills, and other means, we have assessed state and local public health system performance and noted vast differences in the ability to detect a potentially devastating health event and to respond accordingly. When put to the test, some communities demonstrate an impressive level of readiness. However, others have been slow to grasp the potential magnitude of an event, much less initiate a robust response.

Moreover, the poorer-performing communities were frequently unable to provide their residents, including those who are highly vulnerable or at risk for severe consequences, with adequate information on how to care for themselves and others in a health emergency. Upon further analysis, we found that these communities did not actively engage important community groups in preparedness planning and did not fully understand the connections between and among organizations involved in response and recovery. These inequities are in part the result of the federal system; that is, public health is the purview of the state. As a result, the organization of and services provided by public health departments differ across states and even within states. There is no "one size fits all" model of the relative roles and responsibilities of each organization in preparedness and response. Still, the inequities are troubling, and the federal government can play a role in developing strategies to overcome them.

The increased financial stress faced by public health agencies at the state and local levels will certainly constrain their ability to address the challenges described above. With less federal funding through the cooperative agreements and severe state budget cuts, public health agencies are being asked to do more with less. This task is made even more challenging by the structure of federal preparedness funding, which is fragmented, overlapping, and not well coordinated, with funding streams coming from several departments and agencies that are pursuing similar goals. In addition, restrictions on how grant funds can be used by states can impede innovation and creativity in leveraging the funds to increase preparedness. At the same time, challenges in measuring performance have made it difficult to establish how entities that receive public funds will be held accountable for their performance.

## A Path Forward

Since 9/11, attempts to better coordinate preparedness, response, and recovery efforts have moved in fits and starts. The recently issued National Health Security Strategy of the U.S. Department of Health and Human Services provides a useful roadmap and offers a meaningful path forward. Mandated under the Pandemic and All Hazards Preparedness Act of 2006, the strategy delineates two goals and ten key objectives[8] that are required for the nation to achieve "health security." As defined in the strategy, health security is a "state in which the Nation and its people are prepared for, protected from, and resilient in the face of health threats or incidents with potentially negative health consequences." The two overarching goals are to build community resilience and to strengthen public health, medical, and emergency response systems.

While public health preparedness and health security have historically been viewed as the government's responsibility, the national strategy seeks to change this paradigm and highlights the need for all sectors of society to contribute to health-security efforts. The strategy notes that there is much that individuals, community-based organizations, and businesses can do to build community resilience and improve

---

[8] The ten objectives address issues related to informed and empowered individuals and communities, a national health security workforce, integrated and scalable health-care delivery systems, situational awareness, timely and effective communications, effective countermeasures, prevention/mitigation of environmental and other health threats, post-incident health recovery in planning and response, cross-border and global partnerships, and the application of scientific, evaluation, and quality-improvement techniques.

health security.[9] It also recognizes the inextricable link between health security and national security: "Simply put, the health of a nation's people has a direct impact on that nation's security. Any large-scale incident such as a natural disaster or an infectious disease pandemic . . . endangers the security and stability of that society."

The strategy emphasizes the need for individuals and communities to better understand how to care for themselves in both routine and emergency situations—and the need for health care infrastructures to meet both anticipated and unanticipated demands. A detailed implementation plan is also being developed to specify stakeholder roles and responsibilities. Additionally, it will be important to integrate these public health system changes with impending changes in the health delivery system—including those related to implementation of the Patient Protection and Affordable Care Act of 2010.

Widespread adoption of the National Health Security Strategy and its accompanying implementation plan can dramatically upgrade the nation's ability to prepare for, respond to, and recover from large-scale health emergencies. Still, its successful implementation will require a significant cultural shift—one that recognizes that all sectors of society have a stake in, and thus share some responsibility for, health-security planning and execution. Cultivating a sense of shared accountability will require an effective public engagement process. Engaging all relevant stakeholders in discussions about national health security can increase their sense of ownership and thus their willingness to assume responsibility for specific activities and to follow through on them.

A key challenge will be to find the resources needed to merely sustain existing public health capacity, much less improve it. This is an especially difficult task in the current fiscal climate. To ensure that sufficient resources are dedicated to building and maintaining national

---

[9] For example, individuals can develop family preparedness plans, learn basic first aid, and identify neighbors (such as the elderly and disabled) who may need assistance in a disaster. Community-based organizations can play a role in ensuring that the needs of the populations they serve are incorporated into community preparedness plans and in providing information and services during the response and recovery. Private firms can contribute supplies and/or services to a response, take actions to protect their employees, and play an important role in recovery efforts.

health security, it is crucial that existing funding streams be used as efficiently and effectively as possible and that strategies are developed to attract and leverage private sector investments. The return on existing investments can be improved by better coordinating the various funding streams to reduce duplication of effort.

A first step is to inventory existing grant programs across the federal government and identify overlaps and opportunities to better coordinate application, reporting, and performance-measurement requirements. It is also important to evaluate the effectiveness of current and future preparedness programs so that the most promising practices are quickly identified and shared across jurisdictions. New investments could be generated by offering incentives (financial and other) for individuals and institutions to participate in health security activities and efforts.

In the future, it will be important for everyone—from government officials and first responders to directors of nongovernmental community organizations, private sector executives, volunteers, and local citizens—to understand how they can contribute to improving the nation's ability to respond to and recover from the unexpected. Despite progress, there are persistent gaps in core public health services, many vulnerable populations remain vulnerable, and integration between public health and the nation's health-care delivery system is inadequate. Moreover, the combination of a severe recession and the fortunate lack of a major terror attack against the U.S. homeland since 9/11 has fostered a growing sense of complacency among those who are not directly involved. Strong leadership in all sectors of society will be essential to reverse this disturbing dynamic. What the nation needs to do to be prepared to effectively confront public health threats is clear: The challenges going forward are to survive the current austerity, improve performance and accountability, and educate the public and policymakers about the vital role of public health—and their vital roles in safeguarding it.

## Related Reading

Courtney, Brooke, Eric Toner, Richard Waldhorn, Crystal Franco, Kunal Rambhia, Ann Norwood, Thomas V. Inglesby, and Tara O'Toole, "Healthcare Coalitions: The New Foundation for National Healthcare Preparedness and Response for Catastrophic Health Emergencies," *Biosecurity and Bioterrorism: Biodefense Strategy, Practice, and Science,* Vol. 7, No. 2, 2009, pp. 153–163.

Institute of Medicine, *The Future of Public Health,* Washington, D.C.: National Academies Press, 1988.

———, *The Future of the Public's Health in the 21st Century,* Washington, D.C.: National Academies Press, 2002.

Nelson, Christopher, Nicole Lurie, Jeffrey Wasserman, and Sarah Zakowski, "Conceptualizing and Defining Public Health Emergency Preparedness," *American Journal of Public Health,* Vol. 97, Supplement 1, 2007, pp. S9–S11.

Select Bipartisan Committee to Investigate the Preparation for and Response to Hurricane Katrina, *A Failure of Initiative: Final Report of the Select Bipartisan Committee to Investigate the Preparation for and Response to Hurricane Katrina,* Washington, D.C.: U.S. Government Printing Office, 2006.

University of Pittsburgh Medical Center (UPMC), Center for Biosecurity, *The Next Challenge in Healthcare Preparedness: Catastrophic Health Events,* prepared for the U.S. Department of Health and Human Services, 2009.

U.S. Department of Health and Human Services, *National Health Security Strategy,* 2009. As of May 24, 2011:
http://www.phe.gov/Preparedness/planning/authority/nhss/strategy/Pages/default.aspx

CHAPTER FIFTEEN

# The Link Between National Security and Compensation for Terrorism Losses

*Lloyd Dixon, Fred Kipperman, and Robert T. Reville*

*Lloyd Dixon, a RAND senior economist, was the research director of the RAND Center for Terrorism Risk Management Policy (CTRMP). Fred Kipperman, the senior director of strategic relationships for the RAND Institute for Civil Justice, was the associate director of CTRMP. Robert Reville, a RAND labor economist with expertise in compensation and insurance public policy, was co-director of CTRMP. CTRMP closed after completing its planned four-year mission to inform debates related to the Terrorism Risk Insurance Act of 2002 and its extension.*

The focus of much of the nationwide debates on homeland security and foreign policy sparked by 9/11—debates on everything from an Islamic community center near Ground Zero to the war in Iraq—has been on the prevention of future attacks. Largely overlooked in efforts to enhance national security has been a basic question: When Americans are injured or killed or when their property is damaged by terrorists, who, if anyone, should pay to cover the losses? The answer to this question can have important implications for the resilience of the country's economy and society to terrorist attacks. It can also affect America's vulnerability to attacks in the first place. In crafting a national security strategy for terrorism, Americans should think about not only the most effective ways to prevent terrorist attacks but also the programs and policies that provide compensation for the individuals and businesses affected by those attacks that do occur.

Almost everyone agrees that the party responsible for a loss or injury should pay. The responsibility for a terrorist attack lies firmly with the terrorists, their organizations, and their supporters, but the terrorists may be dead, and their organizations are likely beyond the reach of U.S. courts. Without the ability to obtain recourse from the terrorists, on whom will the financial responsibility fall? On the government, acting as representative of the people? On the businesses that failed to prevent the attack or to protect their workers or customers from being in harm's way? On the insurance industry, whose role it is to compensate policyholders for injury? If none of the above, the burden would fall on the victims themselves, but that could exacerbate the economic and psychological effects of the attack and serve the terrorists' interests.

After 9/11, victims received compensation from multiple sources, ranging from charities to an extraordinary federal victim compensation fund. Businesses also received insurance payments, which were the largest for a single-event loss up to that time, as well as government support. It is difficult to determine how much more quickly the economy rebounded than it would have without such programs, but we suspect that the types and amounts of compensation did affect both economic recovery and social cohesion following the attacks.

Unfortunately, there has been no sustained effort to articulate a strategy for the compensation of terrorism losses. Many businesses no longer have insurance that covers terrorist attacks. The Terrorism Risk Insurance Act of 2002 (TRIA), the federal program adopted to increase the availability and reduce the cost of terrorism insurance, is set to expire in 2012. The dire budget environment seems to rule out a generous federal compensation program. Little guidance has emerged from the courts on when private firms can be held responsible for losses caused by terrorist attacks. Nor have we in America considered how the compensation system (or lack of one) may affect the incentives of businesses and individuals to reduce vulnerabilities to future attacks. Consequently, while we may be better prepared to prevent attacks, we may be less prepared to recover.

## Compensation for 9/11 Victims

Following the 9/11 attacks, a massive and largely uncoordinated response that combined insurance payouts, government programs, and charitable contributions provided compensation both to those who suffered injuries or losses in the attacks and to the survivors of the 2,976 people killed either by terrorist action or in the effort to rescue others. A 2004 RAND study, *Compensation for Losses from the 9/11 Attacks*, found that overall compensation to businesses, homeowners, injured individuals, and survivors amounted to more than $38 billion through the time the study was written.[1]

Insurance companies paid more than half of the overall compensation for those affected, and the bulk of the insurance payments went to businesses in New York City, covering property damage and business interruption (see figure below). Those killed or injured at work

**Most of the Compensation Paid to Businesses and Individuals in New York City After the 9/11 Attacks Came from Insurance Companies**

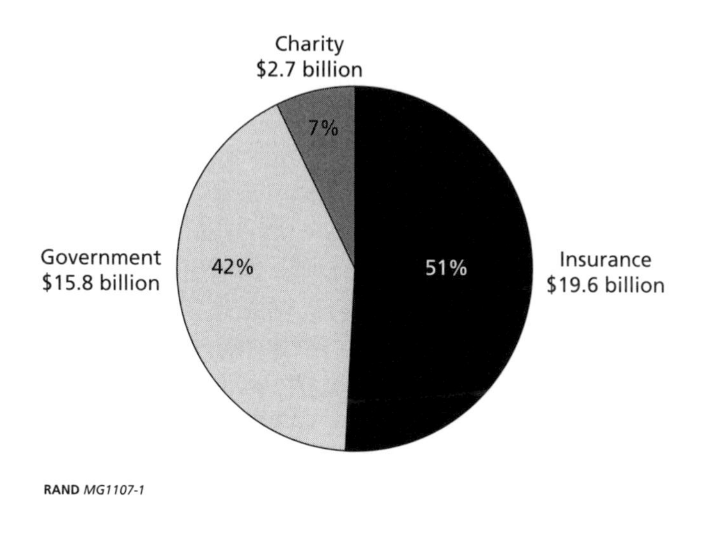


[1] Lloyd Dixon and Rachel Kaganoff Stern, *Compensation for Losses from the 9/11 Attacks*, Santa Monica, Calif.: RAND Corporation, MG-264-ICJ, 2004, p. xviii. As of May 24, 2011: http://www.rand.org/pubs/monographs/MG264.html

were able to collect some benefits provided by workers' compensation insurance and, in some cases, group life insurance payouts. Others were covered by personal life insurance policies.

Probably the best-known means of compensation resulted from congressional action to set up the federal September 11th Victim Compensation Fund of 2001. This fund was designed to do many things: to compensate those who suffered serious physical injury, to compensate the families of those killed in the attacks, to limit litigation, and to provide a visible government response to the unfortunate citizens who arguably were standing in as proxies for all Americans. In total, the fund distributed more than $7 billion to 5,560 claimants. The federal government also provided billions to compensate businesses and workers and to rebuild New York City. The total amount of compensation paid by the federal government was nearly $16 billion.

Nearly two-thirds of U.S. households made contributions to charities for victims of the 9/11 attacks. Total charitable donations that were used for compensation exceeded $2.7 billion.[2] Meanwhile, a very small number of victims opted out of the compensation fund and filed lawsuits against an array of parties, including airlines, property owners, insurance companies, and government entities.

The responses of government, insurers, charities, and plaintiff lawyers, who donated their time to help victims apply to the compensation fund, could be seen as demonstrations of national solidarity against the aims of terrorists. Ken Feinberg, the special master of the fund, has described it as "vengeful philanthropy"—showing the terrorists that they cannot hurt or divide the country, because the United States will support the families of the dead and seriously injured.[3] It can also be argued that the fund itself, by providing relatively rapid compensation

---

[2] Dixon and Stern, 2004, p. 1.

For a detailed evaluation of the performance of the compensation system after 9/11, see the Dixon and Stern volume and also Loren Renz, Elizabeth Cuccaro, and Leslie Marino, *9/11 Relief and Regranting Funds: A Summary Report on Funds Raised and Assistance Provided*, New York: Foundation Center, December 2003.

[3] Kenneth R. Feinberg, "9/11 Victim Compensation Fund: Successes, Failures, and Lessons for Tort Reform," comments at Manhattan Institute Center for Legal Policy Conference, Washington, D.C., January 13, 2005.

and avoiding protracted legal disputes, demonstrated national unity against terrorism and, to some extent, frustrated the ultimate goals of the terrorists.

## Current Compensation Mechanisms for Terrorist Attacks

The benefits paid out after the 9/11 attacks were the result of a unique combination of circumstances that is not likely to be repeated, for an array of reasons. At the time of the attacks, commercial insurance policies did not exclude losses due to terrorist attacks. After 9/11, insurers revised their policies, which now require add-on policies to protect businesses and property owners from losses related to terrorism.

By providing a federal backstop for terrorism-related losses, TRIA increased the availability and reduced the cost of terrorism insurance compared with what would have been the case absent the act, but insurance coverage for terrorism is still not as widespread as it was before 9/11.[4] Moreover, it is not clear that Congress will extend TRIA beyond its 2012 expiration. As a consequence, there is no guarantee that insurance will play a lead role in reimbursing losses after a future attack.

There is also no guarantee that the federal government will step forward with a compensation program as generous as the September 11th Victim Compensation Fund in the event of another large terrorist attack. Feinberg counsels against a similar government program because the fund tied compensation to the income of the deceased or injured party. Instead, he advocates for a federal program that would offer a flat, more modest level of compensation to promote equity, ease of administration, and solidarity.[5] Efforts to expand the fund to cover the 1995 Oklahoma City bombing failed, and in another indication of resistance to this type of federal compensation scheme, it took

---

[4] See Lloyd Dixon, Robert J. Lempert, Tom LaTourrette, and Robert T. Reville, *The Federal Role in Terrorism Insurance: Evaluating Alternatives in an Uncertain World*, Santa Monica, Calif.: RAND Corporation, MG-679-CTRMP, 2007. As of May 24, 2011: http://www.rand.org/pubs/monographs/MG679.html

[5] Kenneth R. Feinberg, *What Is Life Worth? The Unprecedented Effort to Compensate the Victims of 9/11*, New York: Public Affairs, 2005.

many years to reopen the fund to provide compensation to emergency responders and others who had died or fallen ill due to dust inhalation following the 9/11 attacks.[6]

Charity simply cannot be relied upon as a major source of compensation in the wake of a future attack. The state of the economy could reduce giving after an attack, as perhaps would donor fatigue related to the unfortunate regularity of natural disasters around the world. Charitable giving following a new attack would also not be augmented by the shock value of seeing a massive attack on U.S. soil for the first time.

Diminished payouts from insurers, government programs, and charities would mean that the tort system would play a bigger role in compensation following a future attack than it did after 9/11. Unfortunately, the unpredictable tort system is fraught with expense, delay, and lack of meaningful precedent. The absence of legal precedence or clear guidance from government, either through legislation or otherwise, virtually guarantees future legal wrangling about liability and damages following terrorism events.

The September 11th Victim Compensation Fund limited the liability of airlines and a number of other businesses and government entities that had allegedly failed to protect their customers adequately from the consequences of a terrorist attack, but those limitations applied only to the 9/11 attacks. Congress later passed the SAFETY Act of 2002, which limited liability for providers of anti-terrorism technologies.[7] The law is of limited scope, however, and there is significant debate in legal and corporate communities about the extent to which it actually limits liability. Meanwhile, little consideration has been given to how different liability standards might or might not compel businesses to protect their employees and customers.

---

[6] Additional compensation was made available by Public Law 111-347, The James Zadroga 9/11 Health and Compensation Act, January 2, 2011.

[7] The Support Anti-Terrorism by Fostering Effective Technologies Act of 2002 (SAFETY Act) (Subtitle G of Title VIII of Public Law 107-296, The Homeland Security Act of 2002, November 25, 2002). For more information on the act, see the federal Safety Act website (as of May 27, 2011: https://www.safetyact.gov/).

Several post-9/11 developments in law and insurance may have raised more questions than answers regarding the liability for losses caused by terrorist attacks. Lawsuits related to the 9/11 attacks have been filed over site clean-up, foreign sponsorship of terrorism, the duty of property owners to protect their customers and employees, insurance claims, and insurance subrogation.[8] Many of the cases have settled, which on the plus side has resolved the litigation but on the minus side has meant that few legal precedents have been established to clarify the preventive measures expected of businesses that might also be victims of an attack.

Ten years after the 9/11 attacks, the compensation system for losses following a large attack remains undeveloped and highly uncertain. The system we are left with provides clear signals neither on what actions private firms and public agencies should take to mitigate the risk of terrorism to their employees and customers nor on what losses they may be required to cover should another attack occur. The lack of a clear compensation strategy could delay compensation, retard economic recovery, and prolong the visible and psychological imprints of an attack. It could also impose extensive legal and other transaction costs that divert society's resources from more productive uses and distract the attention of business leaders from the recovery of their companies.

## Designing a Strategy for the Future

In his 1998 book *Inside Terrorism*, Georgetown University professor and former RAND counterterrorism expert Bruce Hoffman defined terrorism as the "deliberate creation and exploitation of fear through violence or the threat of violence in the pursuit of political change. . . . Terrorism is specifically designed to have far-reaching psychological effects beyond the immediate victims(s) or object of the terrorist attack. It is meant to instill fear within, and thereby intimidate, a wider 'target

---

[8] *Insurance subrogation* refers to the ability of an insurer that has paid a claim to stand in the shoes of the insured and file suit against a party allegedly responsible for the damages.

audience' that might include a rival ethnic or religious group, an entire country, a national government or political party, or public opinion in general."[9]

Mitigating the consequences of terrorism involves countering this fear. The programs and policies for compensating the victims of attacks may play a role in countering fear as well as in reducing vulnerability to terrorism attacks.

Compensation policies that encourage social cohesion and solidarity could frustrate terrorists' goals. As was seen in the unifying aftermath of 9/11, all levels of government scrambled to offer assistance to businesses and individuals, which undoubtedly reassured citizens and countered the fear stoked by the terrorists. Likewise, Israel's willingness to compensate victims of terrorism in a prompt and equitable manner is seen as a tool to restore life to normalcy as soon as possible after a terrorist attack.[10]

Policymakers should consider a number of factors in crafting a terrorism compensation system. Policies that spread the cost of compensation broadly may further the perception in the United States that a terrorist attack is an attack on the nation as a whole and may thereby reduce fear and promote national unity.[11] Policies that treat all victims similarly—for example, by paying victims a set amount regardless of their economic circumstances—could also promote unity. Although promoting solidarity in these ways may not immediately deter future attacks, it could deter them in the long run by causing terrorism to be less effective in achieving its strategic goals of inciting fear and division.

Compensation policies that reduce vulnerability to terrorist attacks and that help reduce the economic impacts and social divisiveness of attacks that do occur could further dampen terrorist motivations by undercutting the economic costs of terrorism to the country.

---

[9] Bruce Hoffman, *Inside Terrorism*, New York: Columbia University Press, 1998, pp. 43–44.

[10] Organisation for Economic Co-operation and Development (OECD), 2005, *Terrorism Risk Insurance in OECD Countries*, p. 282.

[11] Spreading losses broadly could alternatively encourage resentment in areas where the terrorist threat is low.

Even if compensation policies do not reduce terrorist activity, they can reduce the cascading social and economic consequences of terrorism.

While a tremendous amount of resources and effort has been expended in the name of national security and homeland defense, compensation for the victims of attacks has typically not been considered part of a national strategy for addressing the terrorist threat. This should change. Compensation policy affects both the vulnerability and the resiliency of the U.S. economy and society to a terrorist attack.

After the 9/11 attacks, policymakers limited the liability of businesses, set up the victim compensation fund, and established other economic recovery programs in the heat of the moment, with little chance for careful consideration and analysis. It would be far preferable to work out the roles of the tort system, private insurance, philanthropy, and government compensation and renewal programs in advance of the next attack. Doing so would be an important contribution to our national terrorism strategy.

## Related Reading

"9/11 and Insurance: The Eight Year Anniversary—Insurers Paid Out Nearly $40 Billion," Insurance Information Institute, September 10, 2009. As of June 18, 2011:
http://www.iii.org/press_releases/9-11-and-insurance-the-eight-year-anniversary.html

Chalk, Peter, Bruce Hoffman, Robert T. Reville, and Anna-Britt Kasupski, *Trends in Terrorism: Threats to the United States and the Future of the Terrorism Risk Insurance Act*, Santa Monica, Calif.: RAND Corporation, MG-393-CTRMP, 2005. As of May 24, 2011:
http://www.rand.org/pubs/monographs/MG393.html

Dixon, Lloyd, John Arlington, Stephen J. Carroll, Darius N. Lakdawalla, Robert T. Reville, and David M. Adamson, *Issues and Options for Government Intervention in the Market for Terrorism Insurance*, Santa Monica, Calif.: RAND Corporation, OP-135-ICJ, 2004. As of May 24, 2011:
http://www.rand.org/pubs/occasional_papers/OP135.html

Dixon, Lloyd, and Robert T. Reville, "National Security and Private-Sector Risk Management for Terrorism," in Philip Auerswald, Lewis M. Branscomb, Todd M. La Porte, and Erwann Michel-Kerjan, eds., *Seeds of Disaster, Roots of Response*, Cambridge, UK: Cambridge University Press, 2006, Chapter 18, pp. 292–304.

Hoffman, Bruce, and Anna-Britt Kasupski, *The Victims of Terrorism: An Assessment of Their Influence and Growing Role in Policy, Legislation, and the Private Sector*, Santa Monica, Calif.: RAND Corporation, OP-180-1-CTRMP, 2007. As of May 24, 2011:
http://www.rand.org/pubs/occasional_papers/OP180-1.html

Insurance Information Institute, "The Long Shadow of September 11—Impacts & Implications for Insurers and Reinsurers," March 16, 2009. As of May 24, 2011:
http://www.iii.org/media/presentations/sept11

———, "September 11: One Hundred Minutes of Terror That Changed the Global Insurance Industry Forever," March 13, 2009.

Jackson, Brian A., Lloyd Dixon, and Victoria A. Greenfield, *Economically Targeted Terrorism: A Review of the Literature and a Framework for Considering Defensive Approaches*, Santa Monica, Calif.: RAND Corporation, TR-476-CTRMP, 2007. As of May 24, 2011:
http://www.rand.org/pubs/technical_reports/TR476.html

CHAPTER SIXTEEN

# The Land of the Fearful, or the Home of the Brave?

*Brian Michael Jenkins*

*Brian Michael Jenkins is a senior adviser to the president of the RAND Corporation.*

Testifying before the U.S. Senate in December 2001, when three months had passed with no further terrorist attacks, I was asked if we were "through it yet." I thought not. The United States had just begun the campaign in Afghanistan to crush the group responsible for the most devastating terrorist attacks in history, and in my view it was likely to take years. The question might better be asked today, a decade later. America is now in the tenth year of what was initially called the "Global War on Terror," while the conflict in Afghanistan, a component of that new kind of war, has become America's longest war. The death of al Qaeda's founder and leader Osama bin Laden is a significant blow to al Qaeda and undoubtedly will change the enterprise in ways still unforeseen, but even this event will probably not end its terrorist campaign. We are likely to be at this for years.

How has this experience changed America? With a perspective of ten years, it is possible to discern some effects on how Americans think about personal and national security, see the world, view military power, regard their own government—even how Americans regard each other.

Of course, not all of what Americans have become in the past decade can be chalked up to 9/11 and its aftermath. Growing political partisanship (a trend that preceded 9/11), the country's current economic difficulties, concerns about immigration, America's historically persistent suspicions of internal threats, its obsession with decline, its recurring pessimistic moods, the continuing fascination among many Americans with the end times—all of these independently affect contemporary American attitudes.

Preoccupation with terrorism, including the possibility of terrorism in its worst conceivable forms, has been the most obvious major consequence of 9/11. But terrorism also provided a lightning rod for America's broader anxieties, and it has held a mirror to many of America's enduring characteristics. Had terrorism not become perhaps America's most salient issue, it is hard to imagine it being exploited to arouse concerns about threats to American culture or being blamed for contributing to government deficits and the country's economic decline. It has had deeper effects as well, influencing the legal, social, military, psychological, and even spiritual aspects of life in America. In ten years, the mood of the country has changed profoundly.

## A New Era Defined

The deployment of American forces to wars in the Middle East, Somalia, and the Balkans in the 1990s had lowered expectations of post–Cold War tranquility. But the 9/11 attacks demolished any notion whatsoever that we were at "the end of history."

The post–Cold War era ended sharply on September 11, 2001. History was henceforth to be divided into "pre-9/11" and "post-9/11" eras. Even as globalization facilitated commerce and communication, linking economies and cultures, Americans in the post-9/11 era were more inclined to see the world beyond their borders as a source of danger. As in all threatened societies, perimeters had to be defended, new walls built, vigilance increased. The danger had to be kept outside.

The 9/11 attacks made the response to terrorism the dominant issue in American defense and foreign policy, and the specter of terrorism loomed over American politics, at least through the first half of the ensuing decade (and as a major part of the debate over the invasion of Iraq in 2003). The country reorganized itself for homeland defense, creating the U.S. Department of Homeland Security, now the third largest entity in the federal government. This far-reaching response to 9/11 invited an assertion of executive authority, made security a national preoccupation, and tested the nation's commitment to American values.

## Alarms but No More 9/11s

The first decade of the post-9/11 era was a time of anxiety and alarms. The mysterious anthrax letters, which immediately followed the attacks, raised the specter of bioterrorism. In December 2001, the so-called "shoe bomber" attempted to bring down an American jet flying to the United States from France. To bolster support for invading Iraq, administration officials conflated the terrorist threat posed by al Qaeda with their concerns that Iraq was building weapons of mass destruction. Although, in fact, U.S. intelligence found no link between 9/11 and Iraq, and the intelligence confirming Iraq's possession of weapons of mass destruction turned out to be wrong, the perception of mushroom clouds over American cities remained engraved in the public's mind.

In possession of what it described as "credible intelligence," the federal government raised the terrorist threat level several times between 2002 and 2005, although repetition of these warnings began to prompt public skepticism. Then in 2006, British authorities uncovered an ambitious terrorist plot to bring down a number of commercial airliners flying to the United States from London's Heathrow Airport. Had the plot succeeded, thousands would have died. And in 2009, a Nigerian man with a bomb concealed in his underpants attempted to sabotage a U.S. airliner en route to Detroit from Amsterdam.

Although there were no more 9/11-scale attacks, Americans became concerned about the increasing number of homegrown terrorist plots after an American motivated by al Qaeda's ideology killed one soldier and wounded another at a military recruiting center in Little Rock, Arkansas, and another al Qaeda–inspired gunman killed 13 at Fort Hood, Texas. With these exceptions, there have been no significant terrorist attacks against American targets, either abroad or at home. It is the longest period without a major terrorist attack on the United States since the 1960s.

## The Constitution Holds, Mostly

Faced with the threat of terrorism, most democracies have felt obliged to collect more domestic intelligence, increase police powers, create new anti-terrorist laws, toughen penalties for terrorist-related crimes, and, in some cases, alter trial procedures. Some countries have gone further, imposing censorship on the news media and suspending civil liberties.

Civil libertarians were understandably worried that 9/11 would lead to restrictions on civil liberties characteristic of a police state or, at the very least, would allow a repeat of the abuses that accompanied the domestic intelligence projects of the 1960s and 1970s. Those projects had targeted large numbers of American dissidents, including those who challenged the condition of racial minorities and those who opposed the war in Vietnam.[1] Fortunately, the country did not witness a replay of such programs in response to 9/11. There was pushing and shoving, but the U.S. Constitution held as government steered a middle course.

American presidents made it clear that the war on terrorism was not a war on Islam. Scars left by the unjust internment of Japanese-Americans after Pearl Harbor would have doomed any ideas of internment. It was never seriously contemplated. In 2002, however, the United States began fingerprinting, photographing, and interviewing foreigners arriving in the United States from certain designated Arab and Muslim countries. In order to examine those already in the country before the new measures were implemented, men between the ages

[1] Frank Church, John G. Tower, et al., Senate Select Committee to Study Governmental Operations with Respect to Intelligence Activities ("Church Committee"), *Final Report of the Select Committee to Study Governmental Operations with Respect to Intelligence Activities, Book II: Intelligence Activities and the Rights of Americans*, 1976, p. 22. As of May 27, 2011: http://www.intelligence.senate.gov/churchcommittee.html

of 16 and 45 from these countries and without legal permanent resident or refugee status were required to register with authorities. It was an attempt to uncover the suspected armies of sleeper cells—terrorist infiltrators—that some believe existed in the country.

Reportedly, 85,000 men registered between 2002 and 2003.[2] Deportation proceedings were initiated against more than 13,000, and 2,870 were detained for various violations.[3] How many were deported is unknown; proceedings continue in some cases. Only 11 were suspected of terrorist ties.[4] None were ever charged with terrorist-related crimes.[5] In addition, an unknown number of individuals suspected of involvement in terrorist activity were arrested on lesser charges, and those who lacked citizenship were deported. In the face of widespread complaints about discrimination, this program was scaled back in 2003 and was ultimately suspended in 2011.

Preventive detention of terrorist suspects, a procedure common in a number of democratic countries, was also rejected. The government held Jose Padilla, an American citizen and al Qaeda operative, as an enemy combatant, but his case was ultimately referred to the courts. Americans faced no disappearances, no dungeons.

Seen to have failed, intelligence had to be improved. To ensure better coordination, the intelligence community was completely reorganized for the first time since 1947. At the same time, domestic intelligence efforts were increased, although the U.S. Congress rejected the idea of a separate domestic intelligence service—an American version of Britain's MI5. Instead, the FBI increased the number of Joint Terror-

---

[2] Rachel L. Swarns, "Special Registration for Arab Immigrants Will Reportedly Stop," *New York Times*, November 22, 2003.

[3] American-Arab Anti-Discrimination Committee and Penn State University Dickinson School of Law Center for Immigrants' Rights, *NSEERS: The Consequences of America's Efforts to Secure Its Borders*, March 31, 2009, p. 6. As of June 17, 2011: http://www.adc.org/PDF/nseerspaper.pdf

[4] Sam Dolnick, "A Post-9/11 Registration Effort Ends, but Not Its Effects," *New York Times*, May 30, 2011.

[5] American-Arab Anti-Discrimination Committee and Penn State University Dickinson School of Law Center for Immigrants' Rights, 2009, p. 6.

ism Task Forces created to work with local police, while fusion centers were set up across the country to receive and disseminate information on terrorism passed to them by federal agencies and to collate and analyze information collected at the local level.

Authorities were pushed toward *preventing* terrorist attacks—in police jargon, keeping things to "the left of the boom"—instead of adhering to a traditional, reactive law enforcement approach in which investigation follows crime. Preventive intervention was facilitated both by allowing federal authorities more discretion in opening inquiries in terrorism-related cases and by broadening the crime of providing "material assistance" to a terrorist group, making it easier to investigate and prosecute solely by proving intentions. This was new and troubling territory, although juries accepted it and most prosecutions were successful.

The U.S. Constitution has been severely challenged with respect to the balance of powers. Beginning in 2001, in the immediate aftermath of the 9/11 attacks, the executive branch deliberately bypassed the procedures that had been set up under the Foreign Intelligence Surveillance Act of 1978 (FISA) to ensure that electronic surveillance was judicially authorized. The executive branch claimed in part that the FISA procedures had failed to keep pace with new communications technologies. Congress would have readily agreed to modify the rules to bring them in line with modern technology, but the White House wanted to assert its authority. When the *New York Times* revealed this executive overreach in December 2005, Congress angrily pushed back. With its passage of the FISA Amendments Act of 2008, the long-established requirement of judicial warrant was sustained. But the 2008 law may imply more oversight than is actually going on. The law gave the White House "a significant victory by approving a bill greatly expanding the legal basis for covert government surveillance," according to Judge Richard Stearns of the U.S. District Court of Massachusetts.[6]

---

[6] Richard G. Stearns, "Defining, Detaining, Interrogating, and Trying Terrorist Suspects: The Rise and Implications of a Security State," in Steve Tsang, ed., *Combating Transnational Terrorism: Searching for a New Paradigm*, Santa Barbara, Calif.: PSI Reports, Chapter 2, 2009, pp. 18–33.

The courts would prevail on this count: Suspected terrorists *located in the United States* (including citizens, legal permanent residents, and even foreigners entering the country) would be formally charged and prosecuted in U.S. courts, where independent judges and juries would determine their guilt or innocence. But foreign combatants and suspects captured *outside of the United States* would be treated differently. Some were secretly apprehended and delivered to their home countries or other countries where they faced torture. Those in U.S. hands were held at secret sites abroad or brought to Guantanamo Bay, a U.S. naval base in Cuba, where they would be held indefinitely until trial in a U.S. court or, for some, a military tribunal.

In 2009, a new president tried to close down the Guantanamo detention center, which had become a symbol of abuse, but Congress remained hostile to moving the detainees to U.S. territory. Guantanamo survives, and it is now accepted that some of those held there may spend the rest of their lives there, although the number of prisoners remaining will diminish to a handful of the most dangerous terrorists. Guantanamo remains among the most troubling legal territories.

The complicated trajectory of this issue reflects the White House's initial attempt to keep terrorists captured abroad beyond the jurisdiction of U.S. courts, an effort subsequently rejected by those courts. It also reflects the unique circumstances of the counterterrorism campaign in which war and law enforcement overlap.

## Little Tolerance for Risk

The 9/11 attacks did not create a "security state," but they created a state preoccupied with its security. Government introduced the foreign-sounding term "homeland security" to describe this mission and expressed its commitment with the creation of a large new department dedicated entirely to protecting the public. A powerful actor with a broad mandate, the Department of Homeland Security would secure borders and ports; ensure the security of all modes of transportation; protect vital infrastructure; keep the public safe from chemical, biological, radiological, and nuclear weapons in the hands of terrorists; and coordinate the federal response to man-made and natural disasters.

Long before 9/11, access-control measures had been put in place at most large commercial buildings, hospitals, universities, and, increasingly, residential complexes in response to crime and, to a lesser extent, terrorism. When these failed to protect occupants, litigators were never far behind to demand payment of damages. America had already been attempting to gradually banish personal risk from all aspects of life. The 9/11 attacks added momentum to this long-term trend.

While the federal government was determined to make citizens more aware of security with the new Department of Homeland Security, its direct assumption of security responsibilities fueled both unreasonable expectations and uncharacteristic passivity. Americans increasingly saw it as the federal government's responsibility to prevent all attacks on America, not just to prevent or deter those from hostile foreign states. Any terrorist attempt would be seen not as a consequence of a war in which civilians were on the front line but as government failure. This sobering shift of responsibility does not bode well for the way people may react to future terrorist attacks. It might have been wiser for the federal government to spread the responsibility more broadly across government, to include more active citizen engagement, and to present terrorism as a threat that cannot be prevented but whose likelihood and impact can both be mitigated.

Despite their increasingly unrealistic expectations of absolute security, Americans have remained characteristically contentious about its delivery. Most people are not directly affected by security measures against terrorism, except when they board an airplane. Passenger screening is for many people their only contact with government security procedures, and it has necessarily become more intrusive and intimate, leading to some push back. The recent public resistance to full-body scanning machines and new pat-down procedures is to some degree exaggerated, but it does suggest that the public is near the edge of tolerance, beyond which those charged with security and those being protected become adversaries. It may also reflect broader hostility toward the federal government, which many see as having become too intrusive in all areas. America at war with terrorists ought not become Americans at war with Washington.

## A Nation Continually at War

America fought four major wars in the 20th century and intervened militarily on numerous other occasions, but since 9/11, it has been engaged in continuous military operations for nearly ten years. After the Vietnam War, many said that America was too casualty-averse to support extended military interventions; this has now changed. The contests of the past decade do not approach the carnage of the two World Wars or the wars in Korea and Vietnam in terms of American casualties, but the duration of the current contests is unprecedented. America's involvement in these armed conflicts is likely to continue for some time despite current timetables for withdrawal.

A nation that once saw war as an unavoidable response to foreign aggression is now willing to accept the preemptive use of military force. A people known for impatience now seem more willing, for the most part, to support open-ended military engagement, if not perpetual war, to prevent terrorist threats from arising on distant frontiers. These military operations are seen as necessary components of protecting the homeland against further terrorist attacks. Without 9/11 and the subsequent specter of unlimited terrorism using weapons of mass destruction as portrayed in the rationale for the Iraq War, it is hard to imagine American public support for what are imperial, if not imperialist, missions.

The conflicts in Afghanistan and Iraq have profoundly affected how Americans view military power. They have accepted that some wars may last a long time, and their confidence in America's overwhelming military superiority to vanquish post–Cold War foes has eroded. The dramatic intelligence success and demonstration of American military prowess displayed in locating and eliminating bin Laden has sharply tempered this dim view. But the difficulties confronted in quelling tenacious insurgencies in Iraq and now Afghanistan have made the public and certainly the political leadership decidedly more cautious about engaging in any new military adventures, as the debate over support for the effort to topple Libyan leader Moammar Kadafi bears testimony.

The conflicts in Afghanistan and Iraq have also affected American military doctrine. Following the debacle of the Vietnam War, the U.S. armed forces were determined to never again find themselves waging a long campaign against insurgents abroad while facing mounting popular opposition at home. No U.S. military institution can be expected to carry out a protracted mission for which it lacks popular support. As a consequence, the hard-earned counterinsurgency lessons of Vietnam were systematically expunged from mainstream military curriculum and doctrine, while a new set of criteria was set forth that virtually precluded engagement in anything other than the military's (and the country's) preferred style of war: fast and "winnable" with modern military technology.

America's armed forces were thus ill prepared for the fighting that erupted in Afghanistan after the Taliban were toppled and in Iraq following the swift defeat of Saddam Hussein's conventional forces. Captains and majors had to painfully relearn how to engage and fight an insurgency, often against the military prejudices and historical ignorance of many of their own senior commanders.

The future of American military planning remains uncertain. Some argue that the armed forces must be prepared for the kind of contests that a new generation of officers has now acquired ten years of experience in waging. Others conclude that the military has been taught once again that America should not try to wage wars for which it is ill-suited and politically unprepared, and that instead of preserving the experience of Afghanistan and Iraq, the armed forces must return to their mainstream mission of defending the country against more conventional, even if nuclear-armed, nation-state foes. This argument will persist as a source of tension in planning for national security.

The emphasis on counterterrorism as war has had at least one additional, lasting effect. A nation that previously rejected assassination of terrorist leaders now accepts the routine use of missiles and Special Operations to kill terrorist leaders. In fact, President Bill Clinton authorized the killing of bin Laden in 1999, justifying it as a matter of national self-defense. Still, it is hard to imagine the recent sea change in public attitude were the campaign against al Qaeda not put into the context of 9/11 and a continuing global *war* on terror, in which the killing of enemy commanders is acceptable, even though they may reside in countries outside conflict zones where American forces are engaged in combat.

## American Values Damaged

Although the rule of law prevailed in the post-9/11 decade, American values were at some level damaged. The most egregious example was the employment of coercive interrogation techniques that were tantamount to torture. The sophistic legal defense of these techniques remains a blight on American history. Fortunately, a courageous Senator John McCain, with firsthand experience as a victim of torture, along with other tough-minded patriots, reminded the country of its time-honored American values in this regard, and torture was officially rejected.

Revelations of systematic prisoner abuse at Abu Ghraib prison were another national disgrace. Worse, these crimes were known and concealed by those whose strategic, if not moral, calculations should have informed them that swift action and punishment of those responsible was the correct course.

## From External to Internal Threats

Though many Americans, including the most liberal-minded, were almost ready to flatten Islam's most holy cities after 9/11, America's tradition of religious tolerance also prevailed. Here, President Bush's lead was critical and positive. The president emphasized that the war on terror was not a war on Islam. American Muslims were not the authors of 9/11, and except for a few individuals, they thoroughly rejected al Qaeda's violent ideology. Apart from a handful of ugly incidents, there was no retaliation against the Muslim community.

Yet toward the end of the decade, fears of another terrorist attack gave way to broader concerns about Islam or even about American Muslims. Or perhaps hostilities previously unvoiced found greater reception. Some Americans began to express concerns that the Muslim religion itself posed a threat to American society. Some of these were Islamophobes, but others, worried by Islamist assertions in Europe, asked, How can a nation founded upon separation of church and state and tolerance for religious diversity assimilate groups whose religion insists that state and religion should be inseparable?

American history is filled with communities of faith that voluntarily separate themselves from the society of others, live in enclaves where they can practice purer forms of their religion, and resist assimilation. As long as their religious practices do not conflict with the law of the land, the choice of social separation is theirs. It is doubtful that such separation is the desire of all but a very small fraction of American Muslims, the vast majority of whom have already successfully assimilated into American society without abandoning their faith or compromising their loyalty to the country.

Contemporary American society also provides numerous examples of individuals who would impose tenets of their religious beliefs on others, arguing that God's law must be obeyed. Muslims have no monopoly on aggressive religious politics.

America has always been an anxious nation, distrusting those in power, tending to see conspiracies where there are none, suspicious of newcomers, worried about disloyalty. There is in American society a deep-rooted tradition of political vigilance that at times borders on paranoia dating back to the 18th century. At times, it is wary of the encroachments and perceived secret plans of those elected to power; at other times, it perceives conspiracies aimed at subverting the American way of life. The 9/11 attacks agitated this constant state of distrust, adding Muslims to the historical company of Freemasons, Catholics, Mormons, Jews, Communists, Irish, Italians, Japanese, and Mexicans, to name just a few of the groups that have endured some form of intolerance here. It is likely to remain another continuing source of tension.

## Home of the Brave

The objective of terrorism is to create terror. The 9/11 attacks certainly created a state of alarm. They made the country very edgy, if not fearful, but they did not bring down the republic, imperil America's democracy, or destroy individual liberties.

Americans struck back, albeit not always at the right target, accepting long campaigns but few constraints on freedoms, painfully relearning old lessons, and realizing the limits of military power. At times, American values were threatened or abandoned, but resilient congressional and judicial institutions, cognizant and at times jealously protective of their roles and prerogatives, have struggled mightily to right the ship of state and to chart a new course for the nation. However, this is an enduring task.

Security remains a dominant preoccupation, accompanied by unrealistic expectations on the one hand, yet cantankerous resistance on the other. Tolerance has worn thin. It is a messy but overall optimistic assessment. In the short run of responding to this new terrorist threat, America might appear to be the land of the fearful; but in the long run, I believe, the home of the brave will prevail.

## Related Reading

Jenkins, Brian Michael, *Countering al Qaeda: An Appreciation of the Situation and Suggestions for Strategy*, Santa Monica, Calif.: RAND Corporation, MR-1620-RC, 2002. As of May 24, 2011: http://www.rand.org/pubs/monograph_reports/MR1620.html

———, "Redefining the Enemy: The World Has Changed, but Our Mindset Has Not," *RAND Review*, Vol. 28, No. 1, Spring 2004, pp. 16–23. As of May 24, 2011: http://www.rand.org/publications/randreview/issues/spring2004.html

———, "True Grit: To Counter Terror, We Must Conquer Our Own Fear," *RAND Review*, Vol. 30, No. 2, Summer 2006, pp. 10–19. As of May 24, 2011: http://www.rand.org/publications/randreview/issues/summer2006.html

———, *Unconquerable Nation: Knowing Our Enemy, Strengthening Ourselves*, Santa Monica, Calif.: RAND Corporation, MG-454-RC, 2006. As of May 24, 2011:
http://www.rand.org/pubs/monographs/MG454.html

———, *Would-Be Warriors: Incidents of Jihadist Terrorist Radicalization in the United States Since September 11, 2001*, Santa Monica, Calif.: RAND Corporation, OP-292-RC, 2010. As of May 24, 2011:
http://www.rand.org/pubs/occasional_papers/OP292.html

# About the Editors

## Brian Michael Jenkins

Brian Michael Jenkins, senior adviser to the president of the RAND Corporation, initiated RAND's research on terrorism in 1972. He is the author of *Unconquerable Nation: Knowing Our Enemy, Strengthening Ourselves* and *Will Terrorists Go Nuclear?*

## John Paul Godges

John Paul Godges is editor-in-chief of *RAND Review*, the flagship magazine of the RAND Corporation. He is the author of *Oh, Beautiful: An American Family in the 20th Century*.